Charles Vickerman

Woolen Spinning

A Text Book for Students in Technical Schools and Colleges and for Skillful Practical

Men in Woolen Mills

Charles Vickerman

Woolen Spinning
A Text Book for Students in Technical Schools and Colleges and for Skillful Practical Men in Woolen Mills

ISBN/EAN: 9783744646222

Printed in Europe, USA, Canada, Australia, Japan

Cover: Foto ©berggeist007 / pixelio.de

More available books at **www.hansebooks.com**

WOOLLEN SPINNING

A TEXT-BOOK FOR STUDENTS IN TECHNICAL
SCHOOLS AND COLLEGES, AND FOR
SKILFUL PRACTICAL MEN IN
WOOLLEN MILLS

BY

CHARLES VICKERMAN

AUTHOR OF 'THE WOOLLEN THREAD,' AND 'NOTES ON CARDING

London

MACMILLAN AND CO.

AND NEW YORK

1894

PREFACE

A TEXT-BOOK on Woollen Spinning suitable for students in Technical Schools and Colleges, as well as for skilful practical men in our woollen mills, has long been much wanted; but the retrograde position into which the woollen industry has drifted during late years accentuates this long-felt want, and calls for a text-book that shall restate the principles that underlie the earlier processes of the manufacture.

Woollen goods were "two-thirds of England's exports" at the beginning of the eighteenth century, and down to the middle of the century it was the leading industry, but it was about this time outstripped in the race by the cotton manufacture. The woollen, however, made good progress during the latter half of the last century, and held its position fairly well up to the middle of the present century, since which it has fallen very considerably into the rear.

It is the object of the following pages to restate in short text-book form the principles that underlie the various processes and operations of the earlier portions of the manufacture, and to assert their importance, from the nature of the material in its raw state onward through every operation up to its being ready for the loom, so as to stimulate those who are young and energetic to give attention to the first principles, and to manifest an intelligent interest in the processes in which they are engaged.

I am very much indebted to Dr. F. H. Bowman, F.R.S.E., F.C.S., of Halifax, for allowing me to copy four of the plates (Figs. 1, 4, 5, and 6), with a condensed description, from his great work on *The Structure of the Wool Fibre*, and to Mr. Edward Stanford for permitting me to copy Professor Archer's table on Wools from the volume on Wool in the British Industries Series.

I am also indebted to my old friend Mr. J. B. Wilkinson of Dudley Hill, Bradford, for looking over the part on Dyeing, and for suggestions and additions thereto ; likewise to Mr. Arnold of Huddersfield for revising the worsted drawing.

I have to acknowledge my obligations to Messrs. Dobson and Barlow, machine makers, of Bolton, for so readily and kindly supplying me with a photograph of the head-stock of the Crompton mule, and to Messrs. Platt Brothers and Co. for figures illustrating the various movements in connection with the working of their "patent nosing motion" and their "patent backing-off-chain tightening motion."

However opinions may differ on many points on the woollen manufacture, one thing is certain, that it will need all the help that the truest science and the most perfect practical skill can afford to raise it from its present low condition.

CHAS. VICKERMAN.

Lindley, Huddersfield,
October 1893.

CONTENTS

PART VIII

PART IX

PART X

PART XI

PART XII

ILLUSTRATIONS

PART I

INTRODUCTION

WOOLLEN spinning is different in principle from the spinning of cotton, worsted, alpaca, mohair, and a great many other threads used in our textile manufactures. The peculiar principle of woollen thread construction has not received its due share of attention during recent years—its distinctive peculiarity has either been overlooked or ignored. The woollen industry stands second amongst the great industries of our country, cotton being first, so that anything that tends in any measure to its improvement and development, or the contrary, is a matter of vital importance, both nationally and individually. The woollen industry is, at least, England's most ancient, if not its greatest industry.

It is the object of this text-book not only to set forth the practice, but to state the scientific principles that are involved in the respective processes as they pass in review. It has been said that practice without theory is blind, and has to grope its way forward in the dark; and with equal truth it may be said that theory without practice is also blind, and has likewise to grope its way forward in the dark. Our object will be to go back to first principles, and unite theory, science, and practice on a well-grounded and solid foundation in respect to the structure and spinning of the woollen thread.

From time immemorial man has used the fleece or wool of the sheep for clothing, either in its manufactured, or its natural state on the skin. All down the long pages of history we find mention of its use, and even still further back, in the prehistoric times of our race, when our ancestors lived in those curious lake dwellings, of late so vividly brought under our notice by some of the Gilchrist Lectures. Wherever we find human remains, there we also find the remains of that docile friend of man, the sheep, the patches or fragments of clothing giving us every reason for concluding that the woollen manufacture is one of the most ancient, if not *the most ancient*, of our textile manufactures. The Book of Job is probably the most ancient of writings in existence, and from it we learn that textile fabrics were then in use, for, lamenting his sad estate, Job mentions the weaver's shuttle, and says, " Let me be condemned if I have ever seen any perish for want of clothing, or any poor without covering ; if his loins have not blessed me, and if he were not warmed with the fleece of my sheep." Many of the nations of antiquity claimed the honour of having invented the arts of spinning and weaving. Pliny gives the palm to the Egyptians, and he says that they put a shuttle into the hands of their Goddess Isis, to signify that she was the inventress of weaving. Mitford in his *History of Greece* says, " Of the arts, Egypt was probably the mother of many, as she was the nurse of most," and, according to Bryant's *Ancient Mythology*, the art of weaving was first practised at Arach, in Babylonia, spread thence to the neighbouring cities, then south to Egypt, and in process of time to Greece, and the countries that afterwards formed the Roman Empire. The Babylonians early attained to great proficiency in the structure of tissues and fabrics, and it is recorded of avaricious Achan that when he saw, amid the spoils of Jericho, a goodly Babylonish garment, in all its richness and gorgeous beauty, he " coveted it and took it." Professor Rawlinson says that the Babylonians wore long flowered loose robes and flowered tunics reaching down to their knees, and in their houses their couches were spread with gorgeous coverlets and their floors with rich carpets—all of which must have necessitated immense labour and skill and great knowledge in the manufacture of textile fabrics. It would appear that our advancement in modern times and advantage over the ancients is not so much in the direction of quality as in that of quantity. The wave rolled onward westward, and, by a strange irony of fate, Babylon, the great city of manufacturers and merchants, became desolate, and the very site where it stood has become a moot-point for the antiquary.

The woollen manufacture is of such ancient date that its early progress is veiled in much obscurity, and it is difficult to fix the precise time at which it reached our shores. Some authorities say that a knowledge of textile manufactures was brought to us by the Greeks, and others that we derived our knowledge of their manufacture from those ancient sea-rovers, the Phœnicians, who used to visit the Scilly Islands and the Cornish coasts of Britain to trade for tin.

When Julius Cæsar invaded our island with his Roman legions he found the inhabitants of the southern portion well acquainted with the spinning and weaving of both flax and wool. Two kinds of cloth were manufactured at this period, and both were highly prized by the invaders—one a thick, harsh cloth, which was worn as a sort of mantle, the other of finer wool, dyed different colours, and woven chequered, after the manner of our Highland tartans.   Pliny says the ancient Britons had a method of dyeing purple, scarlet, blue, and many other colours with certain herbs. One plant they used freely for dyeing, and that was the glastum or woad plant.   They appear to have readily picked up the dyeing of many colours, but more especially blue, from their former use of these plants in staining their bodies.

The woollen manufacture does not appear to have been practised in the northern portion of our land prior to the Roman invasion.   Wherever the Romans went they carried their arts and manufactures with them.   In order to benefit our island the Emperors were at great pains to discover and procure the best artificers of every description, particularly manufacturers of woollen and linen cloths, and sent them over to this country, forming them into colleges or guilds, endowing them with certain privileges, and placed them under the great officer of the Empire, *the Count of the Sacred Largesses*.   In this manner the first woollen factory was established at *Venta Belgarum* of the ancient Britons, named by the Romans Winchester, about one hundred years after the conquest of the country.   Thus this manufacture was established, and gradually spread northwards with varying fortunes until the break-up of the Roman Empire, of which event we were the first to feel the effects, as the crumbling Roman power drew in its outposts.   On their departure the woollen manufacture suffered greatly, as woollen garments formed almost exclusively the attire of both male and female of every rank. After the downfall of the Roman Empire the woollen manufacture, which had been with all the arts of civilised life involved in temporary ruin, began first to revive in the Low Countries (now Holland and Belgium) about the middle of the tenth century, where it bestowed peculiar opulence, and considerable social distinction and freedom, on the people engaged in its manufacture, for several hundreds of years.

Taking up again the thread of the history of our own country, after the Romans came the Saxons, who, when they first appeared on our coasts, were a very rude, uncultured, uncivilised race, and under the early rule of this people the woollen manufacture became almost extinct, as well as almost every other art and manufacture established by the Romans. After a time, however, the Saxons fused gradually with the original inhabitants, and wisely assimilated themselves to a civilisation much superior to their own, which included a great portion of what remained of that which was purely Roman. With the uniting of the petty kingdoms of the Heptarchy there came a more settled state of society, which revived the desire for elegance of dress, and the woollen manufacture again improved a little, and "became more artistic in design," for we read in Sharon Turner's *Anglo-Saxons* that "not only were the shuttles filled with purple, but with various colours, and were moved here and there among the thick spreading of the threads (of warp), and by the embroidering art they adorn the woven work with various groups of images."

Such was the state of the woollen manufacture when that grand historic character, our Alfred the Great, was made king. He commanded his household and conducted it in a right royal manner, with a firm, judicious, and steady hand. Sharon Turner affords us a glimpse into its interior. He says, "The Saxon ladies were so much accustomed to 'spinning,' that just as we in legal phrase, and by reference to former habits now almost obsolete, term unmarried ladies 'spinsters,' so Alfred in his will, with true application, called the female part of his family 'the spindle side' of his household, and down to this very day in which we live an unmarried lady is called a 'spinster.'" Whitaker in his *History of Manchester* says, "It is recorded that the mother of Alfred the Great was skilled in the spinning of wool, and trained her daughters to the same pursuit." Spinning with distaff and spindle formed the employment, nay, even the recreation of the noblest of the sex, and we read that the daughters of King Edward the Elder employed themselves in spinning, weaving, and embroidery, which, it is said, " were very prudently then to fill up the very large vacuities of an unlettered life with an innocent and reputable employment." Another account says the king " set his sons to school and his daughters he set to wool-work."

At the time of the Norman Conquest we find that spinning and weaving were again making some progress towards becoming a national industry, though far from having recovered from the blow that it received at the fall of the Roman Empire. Distaff and spindle were at this time becoming common through every rank and condition of society, even down to the very bond-woman, who, working under superintendence, spun yarns by distaff and spindle, and wove the yarns into cloths. The monks were artisans also at this period, in fact, they seem to have been the preservers of the industrial arts as well as learning during the dark stormy times and scenes of the Middle Ages, when society was, as it were, on its trial for life. With William the Conqueror came a large influx of Flemish and Norman weavers, who settled in various parts of England, many in the neighbourhood of Norwich; so much was their dexterity admired that one quaint old author says "the art of weaving seemed to be a peculiar gift bestowed upon them by Nature." To this country the Saxons brought their industry, the Northmen their energy, and the Flemings and Normans their skill, taste, and spirit of liberty; the outcome of these various qualities is the composite character of the English nation. As the Flemings and Normans were proverbial for their love of dress, a new impulse was imparted to the manufacture of wool, and from this period sheep-farming and the growth of wool became a great item in the national wealth. The old historian, Matthew of Westminster, says of this period that "all the nations of the world are kept warm by the wool of England, made into cloth by the men of Flanders." Some authorities even go so far as to say that half the wealth of the kingdom at this period was in wool. The early industry of England was almost entirely pastoral, and down to a comparatively recent date it was a grazing country, and its principal staple was wool.

Many and sharp were the encounters between the barons and clergy on the one hand, and the fiery Norman and Plantagenet kings on the other, who used to seize the wool and raise money upon it with which to carry on their continental and Crusade wars. Stubbs in his *Constitutional History*, vol. ii. pages 126-133, presents a vivid picture of these encounters between monarch and people. Edward I. summoned the clergy of both provinces to meet at Westminster on the 21st September 1294, having previously seized all the wool of the merchants; and so imperious, exacting, and inexorable was the monarch that the clergy were dismayed and terrified, and the Dean of St. Paul's died of fright in the king's presence, and they were obliged

to submit to the most exorbitant requisitions.   And when
the money was obtained a constant wrangle had to be
faced on the subject of foreign service.   The encounter
with Earl Marshal was somewhat lively when the council
broke up in dismay, and thirty vassals joined the two
Earls, Bohun and Bigod, and they immediately assembled
a force of 1500 well-armed cavalry to prevent the king's
officers from collecting money and seizing the wool and
other commodities on their lands.   Some allowance must,
however, be made for the king, his patience having been
sorely tried by these refractory barons, who had failed to
appreciate his most statesmanlike measures.   Like all the
produce in Anglo-Norman times, wool was subject to the
king's right of purveyance and the arbitrary tolls imposed by
the reigning monarch ; but in 1306 a statute was enacted
which considerably limited the king's rights, and which
declared that neither he nor his heirs should have any
tollage or aid without the consent of Parliament, and that
none of his officers should take any corn, etc., without the
consent of the owner, nor should a toll be enforced on
wool.   In 1338 Edward III. took a fifteenth of all the
commonalty of his realm in wool, and in the previous year
he had sent the Bishop of Lincoln and the Earls of Suffolk
and Northampton with 1000 sacks of wool into Brabant,
which they sold at £40 per sack, producing for the king
£40,000 of the then money.   During the summer of
1339 the laity granted to the king one-half of their wool
throughout the whole of the realm, a favour which His
Majesty is reported to have *most graciously* received ; from
the clergy he levied the whole of their wool during the
same summer.   Yet, on the other hand, to bring over the
men of Flanders and employ them in working up the
wool of England in England itself was the constant effort

of the wisest of our Plantagenet kings, and they were just
as constantly opposed by the native workmen and masters
in different parts of the country—workmen then *as now*
failing to see their best interests. A constant war was
going on between the statesmen at Westminster and the
citizens of provincial cities upon the question of the
importation of skilled foreign labour. The laws enacted
in favour of the foreign skilled workmen were counteracted
by the civic regulations to their disadvantage, and much
litigation and many frays, broils, and riots were the result.
It was the same in all parts of the country. Ignorant
opposition was the order of the day, and one Thomas
Blanket, one of the three Flemish brothers who were the
chief promoters of cloth-making in Bristol, was ordered to
pay a heavy fine " for having caused various machines for
weaving and making woollen cloths to be set up in his
house, and having hired weavers and other workmen
for this purpose "; the fine was only remitted by a special
injunction of Edward III., who was a staunch patron of
English manufactures, and who nobly trod in the footsteps
of his grandfather, Edward I. The " Charta Mercatoria,"
granted in 1303, is the foundation from which has slowly
risen our yet but partially developed system of Free
Trade.

Fuller in his *Church History of Great Britain*, published
in 1655, speaking of the time immediately following the
accession of Edward III., who began to reign 1327, says :
" The king and State now began to grow sensible of the
great gain the Netherlands got by our English wool, in
memory whereof the Duke of Burgundy not long after
instituted the Order of the Golden Fleece, wherein indeed
the *Fleece* was ours, the *Golden* theirs, so vast their emolu-
ment by the trade of clothing. Our king, therefore,

resolved if possible to reduce the trade to his own country, who as yet were ignorant of the art, as knowing no more what to do with their wool than the sheep that *weare* it, as to any *artificiall* and curious drapery, their *clothes* then being no better than *freezes*, such their coarseness for want of skill in their making. But soon after followed great alterations." Edward III. was anxious that his people should be independent of foreigners, both for clothing and food. He sent agents abroad to induce Flemish artisans to come over and teach the English the arts of spinning and weaving and dyeing the best kinds of cloth. He promised them protection, and made them such liberal offers that many were induced to accept his invitation. The export of English wool was being prohibited, and as the Flemings could not have the wool brought to their looms, they saw it to their interest and advantage to take their looms to the wool. Accordingly, great numbers of clothworkers came over to this country and settled in London, Somerset, Yorkshire, Lancashire, Kendal, and Carlisle. Letters patent were granted to John Kempe, woollen weaver, in 1330 to exercise his art and teach others. He settled at Kendal, where some of his descendants still reside. Two weavers also settled at York, named William and Hankeimus de Brabant.

Though woollen garments were now extensively worn, the labouring classes were differently clothed, for we read that so late as the reign of Charles I., George Fox, the founder of the Quakers, travelled as a missionary through the country, buttoned up in a leathern doublet with sleeves, instead of a cloth coat, this being the common dress at the time of the labouring mechanics, the class to which this eccentric but gifted individual belonged.

The spinning of wool by distaff and spindle continued in use for some two hundred years after this time; indeed, it was the only known mode of spinning in this country down to the year 1530, when one Jurgens (a baker of Brunswick) invented the one-thread wheel, though there is ample proof for stating that the one-thread spinning wheel was in use in India in spinning cotton long previous to this date. The one-thread wheel may be best described in passing by pointing to the hand-loom weaver's *bobbin wheel*, an illustration of which is given on page 199.

We must not allow ourselves to fall into the error of thinking that there have been no clever people in the world before our day; there were clever people even in those early days, especially amongst ladies. It is on record in the *Transactions* of the Royal Society that a Norfolk lady, of the name of Mary Pringle, earned the notice of the Royal Society by spinning a pound of wool into 84,000 yards (being nearly 48 miles). This clever feat was far surpassed by the achievements of a Lincolnshire spinster, a Miss Ives of Spalding, who spun a pound of wool into 168,000 yards, being 95½ miles of yarn, while the ordinary spinsters of the period reached 13,000 yards for good quality, and 39,000 yards for superfine. There have been, and are, even in our day, some extra-ordinary spinsters amongst the women of Hindustan. The Indian spinster's sense of touch is most acute, and delicate to the very highest degree; she keeps her fingers dry and smooth during the operation by the use of a chalky powder, and in this most primitive manner produces yarns which are much finer and far more tenacious than any of the machine-spun yarns of Europe, and it is from these spinster yarns that are woven those exquisitely beautiful Indian muslins of world-wide fame.

In our own country the one-thread wheel was found in almost every household, for in a process so slow, tedious, and primitive many hands had to be employed, and all the spare moments and breaks in household duties were utilised at the wheel. In some instances as many as 200 spinsters were gathered together in a sort of factory. A wealthy manufacturer in the reign of Henry VIII., of the name of John Whitcomb, had the above number under one roof spinning wool. The one-thread wheel was our best and most expeditious mode of spinning until Hargreaves invented the spinning jenny, or what is now termed the old hand jenny. True, there were two spindles attached to the wheel, but the two spindles never came into general use, except in the case of the Saxony wheel in the spinning of flax. In some of the remote districts the one-thread wheel has had a long existence, and has retained its foot-hold even down to our own times, for in the Exhibition of 1871 there was exhibited from Shetland a hank of wool yarn spun on the old wheel containing 1200 yards, which only weighed $\frac{3}{4}$ of an ounce.

In 1738 commenced the series of changes which has led to our present mode of spinning, when John Wyatt of Birmingham took out his patent for spinning cotton by machinery. This was followed in 1743 by John Kaye inventing the fly-shuttle. In 1769 his son, Robert Kaye, invented the rising-box, or, as it was then called, the DROP-BOX. About the year 1767 James Hargreaves invented the spinning jenny. Before this period one person could attend to only one spindle, and spin one thread at a time. By means of the jenny one person could work twenty or thirty threads at once. Hargreaves increased the spindle power very greatly. He first began with eight spindles in his jenny, and afterwards got as far as eighty spindles. This increase of spinning power caused a riot, and Hargreaves had his spinning machine destroyed by a Blackburn mob, and the poor inventor had to fly from Blackburn to Nottingham, where he sold his invention to the Strutt family, who made their fortunes by it. Hargreaves survived this cruel treatment only for a short time, and died in want and distress, unhonoured and unmourned by the world he had so much benefited, and which rewarded him with cruelty, coldness, and neglect. When the riotous spirit and excitement had somewhat subsided, Hargreaves's jenny speedily came into use, as one person could do from twenty to fifty times as much work as before.

In 1769 Arkwright is said to have invented the throstle frame for spinning cotton; but he did not invent it, he merely took Wyatt's machine and put it into better mechanical form. Arkwright was an adept at seizing the salient points in the inventions of others and putting them into working form. Wyatt had failed because he was unable to get his fitting done correctly. Arkwright did not alter the principle of the machine; the principle has been found incapable of improvement up to the present day; only the mechanical detail has been improved. Arkwright, like Hargreaves, was soon denounced as the enemy of the working classes; his mill near Chorley was destroyed by a mob, he had to seek safety in flight, and ultimately found his way to Nottingham, and established a small business in Wool Pack Lane along with Messrs. Reed and Strutt, who afterwards became his partners. Hargreaves caught sight of one principle of spinning, and Arkwright of the other and opposite principle, as we shall see later on.

Ten years after, in 1779, Samuel Crompton of Bolton

combined the best part of Hargreaves's jenny, the *spindle frame* (and made it movable), with the best part of Arkwright's throstle frame (the rollers), and thus brought out the *spinning mule*, which was ultimately made *self-acting* by Richard Roberts of Manchester, whose QUADRANT WINDING MOTION is one of the most ingenious mechanical movements the world has ever seen. Crompton, having been at work upon his mule for five years, found that he could no longer keep the secret of his machine to himself, and offered to make the principle of the structure of his machine public, providing a subscription was raised in his behalf. The idea of making the thing public on this condition was communicated to the leading manufacturers, and all the money that was subscribed did not amount to more than £106. It is said that Sir Robert Peel, then only Mr. Peel, the father of the statesman, visited Crompton's residence along with two of his men, and minutely inspected and measured the different parts of the new machine, and graciously intimated that the great firm of Peel, Yates, and Company would subscribe one guinea to the fund that was being raised, and in addition gave Crompton sixpence each for the two men as they withdrew for being allowed to take the measurements. After this Peel, Yates, and Company could make a spinning mule as well as Crompton, and the thing was such a success that Peel offered Crompton a situation of trust under the firm, and afterwards a partnership, but Crompton, like most inventors, lacked worldly wisdom, and while smarting under the injustice that had been done to him, declined both offers, and thus lost the only chance he ever had of making money by his invention. Crompton afterwards presented two petitions, at two separate times, to Parliament; and after dancing attendance in the lobbies of the House of Commons, and

after having hope long deferred, at last got a grant of
£5000 from Parliament, which about recouped him the
expense he had been put to, and had to return with a heavy
heart to Bolton, with really nothing for his spinning
mule. Roberts died in the deepest poverty, if not in the
workhouse. If there is one page in Britain's history that
burns with shame it is surely that on which is recorded her
treatment of her inventors. Her patent laws to-day are
the worst in the world, as well as the most expensive:
they do not protect the inventor; they merely give him
the privilege of going into court to protect himself, at his
own risk and cost, instead of its being done for him by a
public prosecutor. The inventor, having paid for a patent,
has a right to have it protected at the public cost.

In 1784 Dr. Cartwright invented the power loom, but
had to lay it aside for many years on account of the
popular hostility towards the further introduction of
machinery. Yet, notwithstanding opposition, the great
work of mechanical improvement still went on, and as one
leader fell another stepped forward, and the battle against
ignorance and technical incapacity continued. Other
obstacles, social, economical, and legal, impeded the progress of the woollen industry, and in the year 1792 an
abstract was published "of laws relating to the growers of
wool, and to the manufacturers of, and dealers in, all
sorts of woollen commodities," and that abstract enumer-
ates and gives the titles of 311 laws on those subjects
then on the statute-book. An Act was passed in the
reign of Edward VI. against the use of machinery in
woollen manufacture, and it was only repealed in
1807. Surely there never was an industry so *be-lawed*
and *statute-booked* in the world before; and such was
the depressing and retarding effect of all this meddling,

that, incredible as it may appear at this day, the same machines or implements were used for carding and spinning in the early years of George III.'s reign as were used during the reign of Edward III., which in all probability were similar to those of the ancient Romans, but more rude in construction. From the most remote period of the woollen manufacture until the latter end of the last century very few, if any, improvements had been introduced into it; with the era of invention that then set in, the woollen, like the other textile industries, greatly benefited. In spite of the restrictions that were placed upon it, the forward movement still went on, and the various changes and improvements operated so powerfully upon the manufacture that in the year 1800 it was found that it had increased threefold in a comparatively short period of time.

Our introductory sketch has now brought us to the beginning of the present century, and within touch of many of the machines and modes of working now prevailing. With these and present-day improvements, alterations, and changes we shall have to deal in detail in the body of this text-book on woollen spinning, in which we shall endeavour to unite practice with the science which underlies the various technical processes through which the wool has to pass before it reaches the loom, so that the work of the hand and the work of the head may grow and flourish together. The limited sense in which the term "spinning" is used includes only the last operation, or finishing process of the series; but it is with spinning in the broad and general sense of the term, including every transforming process that the material has to go through, from the raw state in which the wool leaves the sheep's back, through every operation, up to the point where it becomes a complete textile thread ready for the weaving department, that this text-book deals.

PART II

WOOL

THE wool industries consume a great variety of material in their various manufactures; indeed, so great is the variety that the wools of commerce include in their category almost all the animal downs, such as alpaca, llama, vicuna, mohair, camel hair, the downs of the rabbit and beaver, furs of various kinds, as well as *wool proper*. But with the above animal products other than true wool, the produce of the sheep, we shall only deal incidentally, and in so far as they may serve for illustration.

Unlike many other raw textile materials, such as cotton and silk, which are the product of certain latitudes only, wool is more or less the product of all countries, for no animal so speedily adapts itself to the diversities of climate and pasturage as the sheep, thriving in almost every region where the human race can exist, or is so susceptible of being varied and influenced by the culture and fostering care of man. Such being the case, no single article of commerce is the object of so many dealings as wool, not only as between country and country, but inasmuch as the quality and character of the produce of a single animal varies so widely in the different parts of the fleece as to necessitate its several varieties being used in different sections of the wool industries. Preference is usually given for clothing purposes to fabrics made of wool, so that the extent of their use appears to be only limited by their cost, as they are better adapted to meet the variations of climate and temperature than fabrics of vegetable fibre. Then again, the animal downs differ from the vegetable downs or fibres in possessing in a greater or lesser degree the power of "felting" or "milling," but wool pre-eminently possesses this property, and it is that class of fabrics that are made of wool from yarns peculiarly constructed with a view to their being "felted," "fulled," or "milled" that, properly speaking, in the main constitutes the woollen manufacture.

The rapid strides that are being made by our competitors on the continent and in the United States of America, and the readiness with which they are calling into requisition all that the resources of modern scientific discovery and mechanical invention have placed within their reach, ought to stimulate us to renewed exertion. I am persuaded we are not one jot behind, either in our intellectual attainments or energy and determination ; and I believe that if, instead of resting satisfied with our present position, we keep abreast of the times, and carry into our manufactories and workshops the knowledge which a sound technical education can impart, we shall be enabled to take the lead and maintain our supremacy, whether it be in articles of utility or taste.

Fortunately for us, our rivals, at any rate so far as textile manufactures are concerned, have to work with the same materials as ourselves ; our insular position and our being situated in the very centre of the habitable parts of the globe, and the further fact that we are the great ocean-carriers of the world, makes our country the great central emporium through which the raw materials for the world's use must pass, thus giving us facilities for selection and comparison which are enjoyed by no other nation.

In addition to this we must make ourselves thorough masters of the great principles and laws which underlie all the processes and reactions which the raw material undergoes while it is being transformed either mechanically or chemically into the finished condition. Unless we do this we can never expect to obtain the best results, because we shall be sure to treat the raw material either too little or too much, or subject it to processes which are either unnecessary or unfit for obtaining the object which we have in view. To gain this end it is essential that at the very outset of all our textile manufactures we should have a clear and distinct knowledge of the true nature of the raw material upon which we have to work. No comprehension of general principles can obviate the necessity for this, because this alone can enable us to select the raw material which will best subserve the purpose which we have in view, and enable us to select our various transforming processes, so as to suit the raw material by bringing into play its peculiar properties without injury or detriment to its structure.

The neglect of these precautions, which has in most instances arisen from want of knowledge, has in the past been the cause of very great annoyance and pecuniary loss to the trade. We must all be familiar with numerous cases where the lustre and other properties of the wool have been destroyed by the chemical means employed to cleanse it, or the staple broken and destroyed by the imperfect construction or setting of machinery. Nor is this all, for it frequently happens that the means employed in some of the earlier stages of the manufacture are absolutely detrimental to those which are to follow, and render it quite impossible to attain the desired results. We all know that the spinner and weaver, who are usually quite distinct from dyer and finisher, very seldom either know or consider the processes to which the yarn or goods will be subjected during the dyeing and finishing, and hence it frequently happens that the latter has to remove defects which might be avoided by more care and forethought in the earlier stages of manufacture. On this point I need only instance such a case as the influence of temperature upon the wool fibre, which we shall presently see is a very important point in determining both the after strength, lustre, and softness of the wool, as well as the power which the fibre possesses to receive the dye. We also know how seldom the thermometer is called into requisition during the washing process to determine the heat of the water. Usually it is considered quite sufficient to guess the heat by the immersion of the hand, or the haphazard turning in of steam—a process which, under two different conditions of the body, or by two different individuals, will not be the same within a very much wider range than we should imagine it possible without trying the experiment.

What we want is an intelligent understanding of every process, and the co-ordination of each to the after treatment of the fibre, so that every step may be a step in the right direction, and each process, while perfectly fulfilling its special function, may not in any way interfere with any operation which succeeds it. It may seem at first sight as if this were a comparatively easy matter, and one which a little practice might enable us to settle in such a way that very few mistakes would be made. Experience, however, teaches us otherwise. We must always remember that we can only attain the greatest possible perfection when we have the best possible result obtained in every process through which the fibre passes, and when every stage in the onward course is arranged and carried out with the final consummation in view. This perfection we can only attain by a thorough technical knowledge of the special work which each process is intended to perform, and this, in turn, will always depend upon our knowledge of the structure and affinities of the raw materials which we use. It is necessary, therefore, in order successfully to use each raw material for its best purpose, and in its best way, that we should make ourselves thoroughly acquainted with its typical character and special structure, both mechanically and chemically, and when this is done we can then enter into the principles of the transforming processes with an accurate knowledge of what we wish to produce, and a reasonable expectation of being enabled to accomplish it. The subject of the structure of the wool fibre is more difficult than that of cotton, because the structural differences between the various classes of wool are more varied than those of different classes of cotton, and the effects of climate and species are more marked than in the vegetable kingdom. Chemically, too, the difficulties have been greater because there is a greater vital action in the sheep than in the cotton plant, and the composition of the wool is more affected by the state of health of the animal and the food upon which it is kept.

Hair, wool, and even the feathers of birds are similar to each other in their essential nature, and are all produced in the same manner by an increased production of epidermic or skin cells at the bottom of a flask-shaped follicle. This follicle is formed in the substance of the true skin, and is supplied with an abundance of blood by a special distribution of vessels in its walls. These vessels are continued as a fine network a short distance beyond the root, and thus feed the cells till they are fully developed ; and in the case of some diseases of the hair, such as *pica polonica*, they become enlarged and allow the blood to penetrate up into the substance of the hair, so that if the hair is cut or broken it bleeds. If we pull a hair out by the root we find a bulbous enlargement at the bottom end, of which the exterior is tolerably firm, but the interior consists of a substance or pulp. The continual production of this pulp in the bottom of the follicle, and its conversion into the substance of the hair as it is pushed upwards to the surface of the skin, is the cause of the growth of the hair; but it gives no explanation of the differences which are manifested in the special forms which different hairs and wools assume on different animals, although the appearance of the bulbous parts is very similar. All the parts are only modifications of the various parts of the skin, which have had a vertical rather than a horizontal determination given to them. This is probably occasioned by the presence of nerve fibres, which penetrate the outer sheath of hair within the follicle at its lower part, and thus the hair is put in direct connection with the nervous system. A sudden shock to the nerves will sometimes cause the erection to be so great as to warrant the expression that the hair stands on end.

At first sight it might appear that the skin is a very simple membrane which covers the whole exterior of the animal body, and that the hair or wool simply grows upon the surface in the same way that the cotton fibre grows from the outer layer of the seed wall, and that it possesses some root-like attachment which enables the hair to remain fixed to it. The skin is, however, a much more complicated structure, and it is quite necessary for us to understand something of this structure before we are able to investigate the formation, growth, and nature of the wool fibre.

The skin of all the higher vertebrate animals consists of four strata or layers. Two of these layers, called respectively the *cuticle* or scarf-skin, and beneath it the *rete mucosum*, lie on the outside, and form what is called the *epidermis*. This epidermis serves as a protection to two deeper-seated under layers, which are called the *superficial* or *papillary* layer and the deep-seated *stratum* or *corium*. These two together are called the *dermis* or true skin. We may tabulate these various strata as follows, beginning at the outside—

1. Cuticle or scarf-skin   } Forming the epidermis.
2. Rete mucosum   . }
3. Papillary layer   . } Forming the dermis.
4. Corium   .   . }

If we take a brush, or any rough substance, such as a piece of cloth, and rub it upon the surface of our skin, especially in a dry place, we shall find a quantity of dry dust fly off and attach itself to the cloth. Upon examining this under the microscope we find that this dust is really composed of a series of flattened scales or cells. These are called epidermal cells, and the whole of the skin which lies upon the surface is entirely composed of these microscopical scales, which are really dried-up and dead cells produced from a series of more rounded cells, which lie immediately beneath the flattened ones. These scales form the cuticle or scarf-skin. The rete mucosum is a series of much more rounded cells, which have a distinct nucleus or germinating point, and lies conformably upon the papillary layer or dermis. The density and hardness of the epidermis decreases as we pass inwards from the outer surface to the rete mucosum. This difference in the density is dependent on the mode of the growth of the epidermis, for as the external surface is constantly subjected to destruction from attrition and chemical action, so the membrane is continually reproduced on its internal surface, new layers being constantly and successively formed on the derma to take the place of the old ones. The papillary layer of the derma which lies immediately below the rete mucosum and the upper surface consists of a series of conical prominences or papillæ, which pass upwards into the epidermal layer above it. On the general surface of the body the papillæ are short and minute, but they increase in size and differ in arrangement in various special situations, as in the human body on the surface of the hands and feet, and other parts where great sensibility is requisite. This layer is chiefly composed of areolo-fibrous tissue, elastic tissue, and smooth muscular fibre, together with blood-vessels, lymphatic vessels, and nerves, and is highly vascular and sensitive.

The corium or deep-seated stratum lies lowest of all. The upper surface, which is connected with the papillary layer, consists of areolo-fibrous tissue, which is collected into cellular bundles called fasciculi. They are small and closely interwoven in the superficial strata, and large and coarse in the deeper strata; in the latter forming an areolar network with large openings. These openings are the channels through which the branches of vessels and nerves find a safe passage to the papillary layer above, in which, and in the superficial strata of the corium, they are chiefly distributed.

The dermal and epidermal layers are everywhere penetrated, from beneath to the surface, by a series of channels or openings called sudoriparous ducts, through which the perspiration or moisture from the fatty or adipose layer, which lies beneath the derma, escapes into

the air. The secreting glands which throw off this
moisture are embedded in the substance of the derma and
the subcutaneous tissue, and present every degree of
complexity, from the simplest follicle to the compound
lobulated gland.

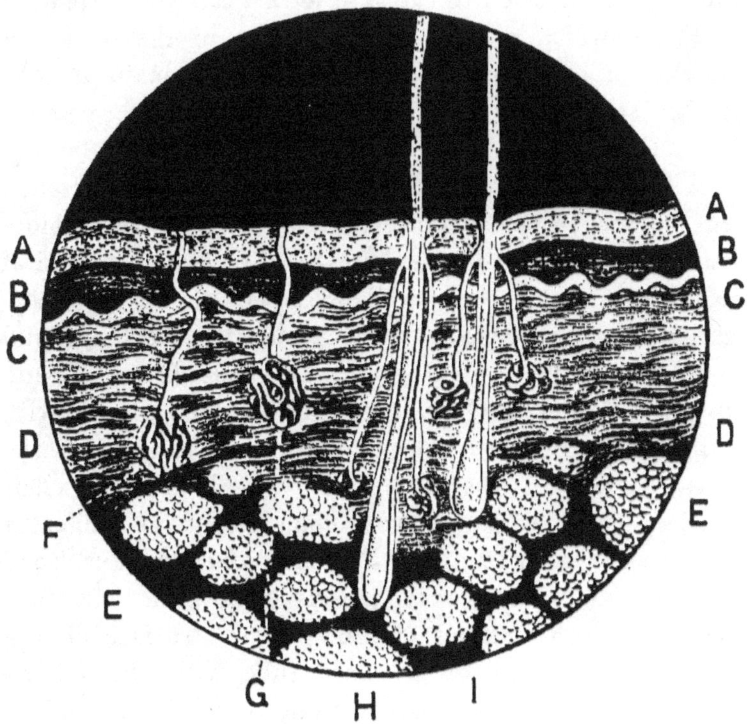

FIG. 1.—SECTION OF SKIN.—25 Diameters.

In some situations the excretory ducts open, as we
have seen, on the surface of the skin, while in others they
terminate in what is of most interest to us, viz. the
follicles of the hairs of wool, which, like themselves, have
their origin usually beneath the skin, and pass upward
through it to the outer surface of the body.

Hairs of wool are therefore living appendages of the skin, produced by the involution and subsequent evolution of the epidermis : the involution constituting the follicle or sac in which the hair is enclosed, and the evolution the shaft of the hair. Before we look at the structure of the hair and its relation to the skin let me call your attention to Fig. 1, which gives a representation of the section of the skin of an animal, from which we can obtain an idea of its structure much better than by any possible description.

A is the cuticle or scarf-skin, with the dead flattened cells on the surface ; B, the rete mucosum, with its more distinct cellular structure, and resting upon the conical prominence of C, the papillary layer. D is the deep-seated strata or corium, with its areolar openings and vascular structure, E the subcutaneous layer of adipose cells. F and G are two sudoriparous glands, with their winding spiral ducts, which terminate on the outer surface of the skin. H and I are two hair follicles, hairs which have their origin at different depths in the subcutaneous tissue, and are furnished on each side with two sebiparous glands, the excretory ducts of which, instead of rising to the surface of the skin, are connected by a short tube with the follicle of the hair. These glands secrete an oily substance, a kind of natural pomatum, with which the surface of the hair is bathed, and thus the scales on the surface are greased and prevented from irritating the nervous lining of the follicle in their passage upwards. It also serves to support and feed the growing hair. The hairs or wool fibres themselves are formed within the follicle in identically the same manner as the epidermis is formed by the papillary layer of the derma. Plastic lymph is, in

the first instance, exuded by the capillary plexus of the
follicle, and the lymph undergoes conversion, first into
granules, and then into cells, which, as the process of
growth and extrusion proceeds, are elongated into fibres,
which form the central structure of the hair. The true
distinction between wool and hair lies in the nature of the
epidermal covering with which the cortical part of the
shaft is covered, and in the method of attachment of the
scaly plates or flattened cells to the inner layer upon
which they rest, and not upon the curly nature of the
fibre itself, although there can be no doubt that this
waved appearance is one of the recognised characters of
wool.

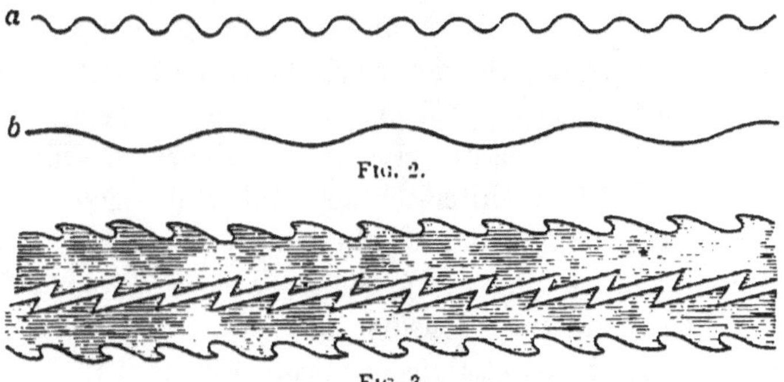

DIAGRAMS SHOWING THE PECULIAR CONSTRUCTION OF THE WOOL FIBRE.

There are, however, to be found numerous hairs which
possess no curl, and yet have the epidermal characteristics
of a true wool, which consists in the power to felt or mat,
arising from the greater looseness of the scaly covering of
the hair, and which, when opposing hairs come into contact, enables these scales to interlock with each other, and
holds them together quite independent of any friction or
twist imparted to the fibre by mechanical means. This
peculiar characteristic will be more readily understood by

looking at Fig. 3, where we have an imaginary section of a typical wool fibre opposed to another similar fibre, so that when they are drawn along over each other the scales interlock, serration into serration, and thus become perfectly united by the wedged edges of the scales entering into the spaces between the scale and shaft of the opposing fibre. We have already seen that in the case of human hair the free edges of the scales are always pointed upwards towards the unattached end of the hair. This is also the case in wool, and when it is in its proper position on the back of the animal, quite independent of other causes which we shall afterwards have to name, the scales of the woolly hair are all pointing in the same direction, so that their tendency to mat or felt is reduced to a minimum, otherwise the fleece or felt of the creature would become one matted tangled mass. The scales being in the same direction, the hairs have the tendency to slide over each other without interlocking, and thus prevent the disagreeable results which would otherwise occur.

If we look at Fig. 4 we shall see there a representation of what we may consider a typical wool fibre. In the wool the cylindrical or cortical part of the fibre is entirely covered with very numerous lorications or scales, the free ends of which have a pointed rather than a rounded form. This enables them when opposed to each other to find their way more readily under the opposing scales, and to penetrate inwards and downwards proportionately to the pressure which is applied to bring them together. In the wool fibre also the free margins of the scales are much longer and deeper than in the hair, where the overlapping scales are attached to the under layer up to the very margin of the scale, which can, even at its extremity, only be detached by the use of a suitable reagent.

In wool this is quite unnecessary, because the ends of the
scales are free to about two-thirds of their length, and are
to a certain extent indeed turned partially outwards, as
can readily be seen by looking at the edges of the wool
fibre, where the denticulated structure is quite distinct
against the dark background.   This   ideal   diagram

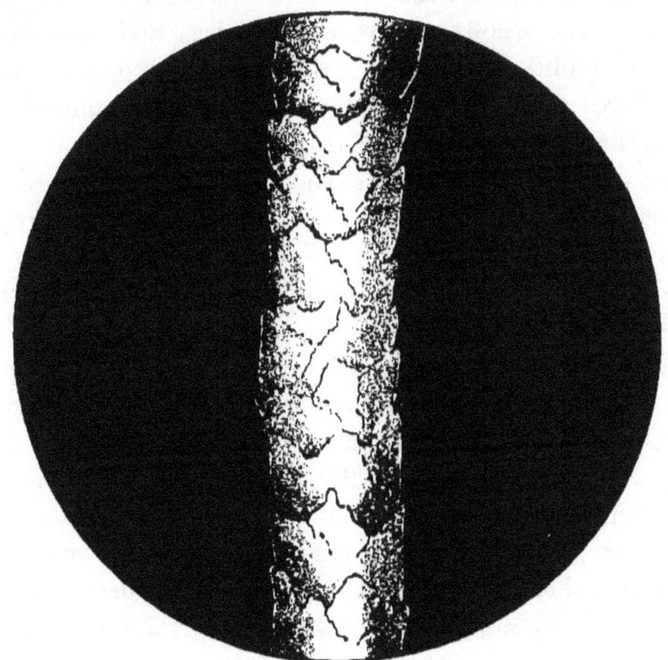

approaches nearer to the fibre obtained from the wool of
the merino sheep than any other, and as we shall presently see this is one of the most valuable and beautiful
wools grown, and one which, either pure or mixed
with other breeds, yields a large proportion of the wool
used in our best manufactures.   This, then, we may
take as our typical wool fibre, and we shall find that

just as the mechanical structure of any wool approaches this standard it becomes better and better fitted for our textile purposes, and so much as it falls short of this standard it becomes proportionately less useful and less fitted for those peculiar uses to which wool can be put. We shall see that in different classes of sheep every variety of intermediate structure occurs between true hairs and the fullest development of the woolly nature of the fibre, and that under certain conditions these wide differences may even exist in the fibres taken from the fleece of the same animal; we shall, however, also see that just in proportion as the conditions surrounding the sheep are favourable to the finest development of all its best qualities, and in proportion as the purity of the breed is maintained, so does the production of hairy fibres diminish and true wool replace them.

However correct our mechanical processes of spinning may be, we can never get beyond the degree of uniformity which is manifested in the raw material itself; this is also equally true as regards the wool fibre. Perfectly uniform material to work upon is the basis of perfect spinning, and although we can never expect to reach this standard, there is no limit assigned to the nearness with which we may approach it. Every step we take towards this end is a step in the right direction, and a knowledge derived from observation and experience of what we require is the only basis upon which we can work. Hence our typical wool fibre becomes invested with a deeply practical as well as theoretical interest, and forms the key to our future progress and success.

The next two figures represent the result of crossing the merino breed of sheep with other choice breeds of longer, more lustrous, and firmer stapled wool. This class of

crossing gives us a wool that approaches very closely to
the highest type of wool fibre, where we have all the best
qualities exhibited. The high-class fibres are all the
result of the cultivation of the merino sheep and its
judicious crossing with the new Leicester and other

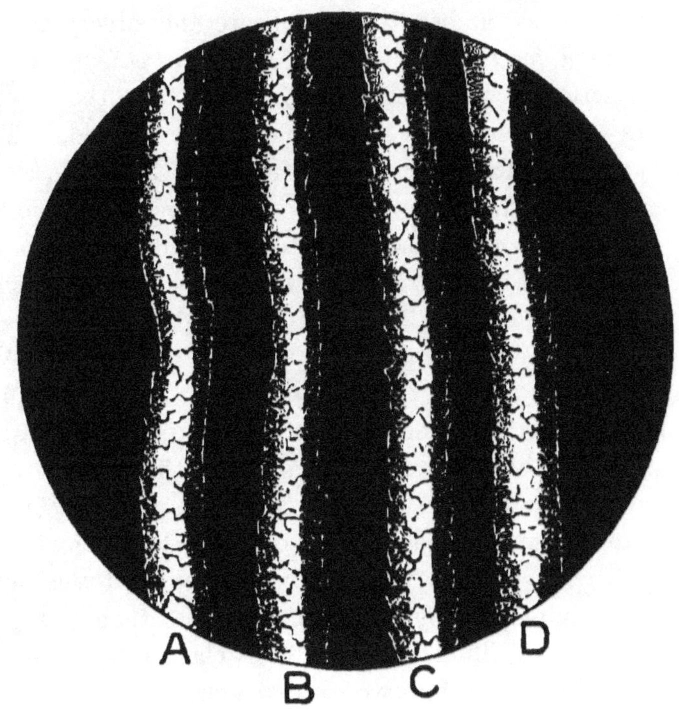

FIG. 5.—MERINO WOOL.—450 Diameters.

first-class long-woolled breeds, so as to increase the
length of the staple and render the wool fit for combing
purposes as well as for carding. We need only give
two illustrations of these fibres. In the first, Fig. 5,
we have a reproduction of a number of fibres of the
finest American merino, which even exceeds in beauty

and fineness of fibre the best Saxony or Australian wool. Some of the fibres were not more than $\frac{1}{30000}$ of an inch, and the scales so numerous that they were no farther apart than half the distance of the diameter of the fibre, so that there must have been about 6000 to an inch. This was the finest wool I ever saw; the whole structure of the fibre was so beautifully delicate and silvery that it was quite a picture to look at. The fibres A and B are the perfection of wool, and even C and D, from the coarser part of the fleece, leave nothing to be desired. The length of the fibre was about 2 inches. Of course when we cross the fine-fibre-woolled with the long-woolled breed of sheep we increase the lustre of the fibre, because we increase the size of the reflecting surfaces which diminishes the dispersion of the light, but we can also increase the diameter of the fibres and their general coarseness. We gain, however, the length of staple which enables the fine wools to be used for worsted spinning, and at the present time such has been the improvement in the wools grown in some of our colonies that all classes of English bright wools can now be replaced by some of the long-woolled foreign varieties, and even where not replaced the yarns are much improved by an admixture with them. Fig. 6 gives a typical illustration of this class of wool—the Leicester Botany. Here we have the curl and softness of the merino united with the length and lustre of the best deep-grown English wool. The diameter of the fibres is reduced from an average of about $\frac{1}{800}$ of an inch, as in the pure Leicester, to $\frac{1}{1000}$ of an inch, and the number of scales per inch also increased in about the same proportion; while, with the purity of the climate and the care and attention bestowed upon the sheep, the fibres become wonderfully uniform, and the lustre bright and silvery, some of them approaching in brightness the lustre of the mohair.[1]

Dr. Bowman very aptly observes that "the future of manufactures seems to me like the future of scientists, to

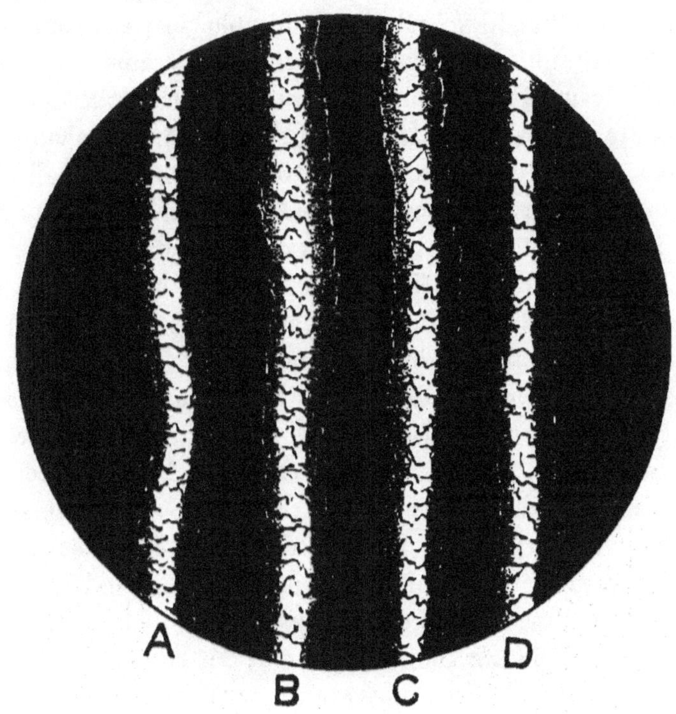

Fig. 6.—AUSTRALIAN WOOL (Botany).—350 Diameters.

point to *specialists*, and we already see this in the fact that certain individuals, and certain districts, obtain a special name, which the necessary adaptation in every

[1] See *The Structure of the Wool Fibre*, by F. H. Bowman, D.Sc., F.R.S.E., F.L.S.   Manchester: Palmer and Howe, Princess Street; London: Simpkin, Marshall, and Co.; Philadelphia: Henry Carey Baird and Co.

respect alone can give; and it will be more so in the future, as a continually higher standard of perfection will be required. What shall I say in regard to the neglect of all consideration in the mechanical processes which is often evinced in respect to after chemical processes, such as dyeing and finishing, where unsuitable soaps, unsuitable oils, unsuitable temperatures, both moist and dry, are frequently used in absolute forgetfulness that the delicate fibres are afterwards intended to receive various shades of colour, and need all the porosity of the fibre, and all the original surface lustre of the epidermal scales, if they are to give full effect to the beauty with which they were originally endowed. Then we wonder how it is that certain goods turn out wrong, and every one concerned blames every one else in a ' vicious circle.' We have now seen how delicate is the structure of the fibre, and how easily the fine enamelled surface of the epidermal scales can be injured, either by too great heat in the water, or the use of alkaline leys, which remove not only the suint and free fats which are associated with it, but also attack the fatty constituents of the structural cells, and thus not only tend to render the fibre hard, unpliable, and hask, but also deteriorate the strength and destroy the lustre."

Comparatively little was known regarding the structure of the wool fibre previous to the series of investigations commenced in 1833 by the late Mr. Youatt, the eminent veterinary surgeon, of London, aided by Mr. Powell, the optician, and the late Mr. Thomas Plint of Leeds. The result was published in 1837 in Mr. Youatt's work on *Sheep : their Breeds, Management, and Diseases.* These investigations were followed by those of Professor Low, Mr. Goss, Dr. Lankester, and many others, but the most recent and thorough investigation in this field of inquiry is that of Dr. F. H. Bowman of Halifax, whose work on the *Structure of the Wool Fibre* is a marvel of patient, painstaking inquiry, and we are greatly indebted to Dr. Bowman's kindness for being allowed to copy the four plates (Figs. 1, 4, 5, 6) on the wool fibre out of a series of over thirty in his great work, and to make extracts from the letter-press descriptions of the same, thereby being enabled to lay before the students of our technical colleges and schools in short text-book form the result of the most recent and most searching investigation into the structure of the wool fibre.

We will now proceed to notice a few more points in relation to wool. Not only is the wool fibre a tube, but even the tube itself is said by some writers on wool to be composed, or built up, of a number of filaments or smaller tubes, ranged side by side, and forming the shell of the tube of the wool fibre. Although wool as well as hair is of tubular construction, yet that cylindrical form varies with the climate in which it is grown. A cross-section of a hair of wool, if strictly circular, denotes that it has been grown in a cold northern climate, and the hair or wool is lank, long, and soft; but if the cross-section is flat-sided or oval in the extreme, the wool is of tropical growth, and is crisp and frizzled. There is a marked improvement in the animal downs as we approach the higher latitudes, and the better class of wools for clothing purposes can only be grown in temperate climates. Too hot a climate yields a wool too hard, crisp, dry, and frizzly; while, on the other hand, too cold a climate, though yielding a wool that is soft to an extreme degree, gives too little of the wavy curl or frizzle for many of the higher class of goods of the woollen manufacture. There are other points about wool the absence of which has great influence with the manufacturer when looking over the raw material with a view to purchase. The manufacturer likes to see trueness of breed all over the fleece, with the staples uniformly full of compact wavy curl, and to find no shaggy portions interspersed, which occurs only in poor neglected flocks. Another point is soundness, and it is only a branch, as it were, of the previous matter, as it is inseparably connected with evenness ; for wherever evenness is lacking, soundness will be lacking too, as both have their rise and origin in *healthy, well-cared-for, well-conditioned sheep.* If the hair is bright and clear, healthy, and moderately yolked, we are almost sure to find soundness ; but where the fibre has a peculiarly dull appearance, and is dry with a withered or wizened aspect, we may expect to find tenderness. The buyer does not trust to the unassisted eye, but picks out here and there a small staple, and takes hold of each end, and pulls until the staple breaks ; this test he applies quickly at intervals, and quite mechanically, as it would seem, while his examination proceeds. Soundness is of great importance to the manufacturer, and he is quite aware that where this is wanting other things will be wanting also. Age has much to do with determining the soundness of the fleece. After the animal has passed a certain age (say six years, especially in the female), the yolk begins to lessen, and with it will lessen also the soundness and healthy character of the wool. As the yolk lessens there is a tendency in the fleece to "cotting" or felting, a thing which it cannot do so long as the yolk is in healthy and vigorous flow. Wherever the wool is found to be "cotted" it is a sure sign that the sheep are either old or unhealthy ; in either case the prudent manufacturer holds aloof. A further point intimately connected with the above is softness. Where the two points of evenness and soundness are found we are pretty sure to find softness in close company with them. It will now be seen that it is the softness of healthiness that is more particularly implied here and required, for softness apart from soundness is of little value to the manufacturer. Softness in a great measure depends upon the fineness of the fibre, though the kind of breed of sheep has much to do with the point also. It also depends like soundness upon the presence and quantity of the yolk, for where this is wanting we can neither obtain soundness nor softness in the wool. The sheep must be cared for, and, in fact, cultivated up to the condition at which the yolk flows freely. After softness comes fineness, and this characteristic is not altogether confined to short wools, as some of the long wools are not only soft but fine also—a rare combination. Not only have we fineness in the abstract in regard to the fleece as a whole, but the wool varies in a very considerable degree between one part of the fleece and another; some portions of the same fleece are five times as fine as other portions (see Fig. 7, p. 68), and tested by weight, after being thoroughly cleaned, the portion from the extremity (taken length for length) is often found to be twenty-five times as heavy as the same length taken from the centre or fore-flank, but there is considerable variation in this respect in different breeds. Our finest wool for the superfine west of England clothing purposes comes from Silesia, and is the product of a sub-variety of the merino, which is the finest-woolled breed of sheep in the world. The breed was transferred from Spain to Germany; but Spain was not their original home; they had been carried thither from Greece by the Romans, and during the anarchy which followed the break-up of the Roman Empire, they were in danger of becoming extinct, and were only preserved in the remote valleys and little-known districts of Spain.

The following table, compiled by Professor Archer for his article on " Wool and its Applications " in the series of volumes on " British Industries," which, by the kindness of Mr. Edward Stanford, we are permitted to copy in illustration of the variety of wools to be found in the stocks of the dealers at the present day, is probably the best and most carefully prepared of any that has hitherto appeared, and will give the reader a more correct idea of the numerous kinds of sheep, and the qualities of their fleeces, than might otherwise be obtained :—

Varieties and Sub-varieties.	Breed.	Cross.	Staple.	Quality.	General Colour.	Average Weight of Fleece washed.	Whether for Woollen or Worsted.	General Application.
1. Spanish (Ovis Hispanium of Linnaeus)	Spanish Class 1.—Estantes or stationary		Short	Fine	Black and white	4 to 5 lbs.	Woollen	Used in Leeds and Huddersfield. Spanish wools obtained from the plains are of the merino kind, and are chiefly used for woollen goods, but that obtained from the mountains is coarse and of unequal quality, and is used for low-class goods of various kinds.
	a. Churrah		Long 8 inches	Rather coarse	White	..	Worsted	
	b. Merino		Short	Very fine	White	Ram 8 lbs. Ewe 5 lbs.	Woollen	
	Class 2.—Transhumantes or migratory							
	a. Leonese nigrettes		Short	Fine	Black, white, and gray		Woollen	
	Escurial or Estremadura		Short	Finest	White		Woollen	
	Guadeloup		Short	Very fine	White		Woollen	
	Paulars		Short	Good	White		Woollen	
	Infantados		Short	Coarse and hairy	White			
	b. Sorian							
	Swedish	Merino and native	Long	Soft and fine	White			
	French	Merino and Rousillon	Long	Soft and very fine	White	9 lbs.		
	Danish	Leonese and native	Medium	Fine	White		Worsted and woollen	
	Saxony	Merino and best native	Short	Finest	White		Worsted and woollen	
	Prussian	Merino and native	Short	Very fine	White		Worsted and woollen	

Table showing the Varieties and Breeds of Sheep, Foreign and British—*Continued.*

Varieties and Sub-varieties.	Breed.	Cross.	Staple.	Quality.	General Colour.	Average Weight of Fleece washed.	Whether for Woollen or Worsted.	General Application.
1. Spanish — con-tinued	Class 2 — con-tinued							
	Silesian	Merino and native	Short	Very fine	White.		Worsted and woollen	Silesian wool is almost if not altogether the finest in the world.
	Hungarian	Merino and native	Short	Fine	White		Woollen	
	Hanoverian	Merino and small native	Short	Very fine	White	Ram 4 lbs. Ewe 2½ lbs.		
	New South Wales	Merino and Southdown		Fine	White	2½ lbs.	Worsted and woollen	
	New South Wales	Merino and Leicester		Fine	White	3 lbs.	Worsted and woollen	
	Western Aus-tralia	Merino and Leicester						
	British (pure breed)							
	British	Merino and Southdown		Fine	White			
	..	Merino and Leicester		Fine	White			
		Merino and other native breeds		Fine	White			
2. Common sheep (Ovis rusticus of Linnæus)								
Sub-variety (a) Hornless or Lin-colnshire	Lincolnshire	Lincoln and Leicester	Long	Good and glossy	White	8 to 9 lbs.	Worsted	These are amongst the finest of the long-stapled or worsted wools.
Sub-variety (b) Muggs and Shet-	Shetland		Long	Very fine			Worsted	

Sub-variety (c) Ryeland	Herefordshire	..	Long	Medium	White	6 to 7 lbs.	Worsted	
Sub-variety (d) Southdown	Sussex	..	Short	Fine	White and gray	3 to 4 lbs.	Worsted and woollen	Hoggett's are valuable, the long qualities are used in Bradford and the shorter ones in Rochdale for flannels.
	Kent	Southdown and Romney Marsh	Short	Medium	White	3 to 4 lbs.	Worsted and woollen	
	Hampshire	Southdown and black-faced Berkshire	Short	Fine	White	4 lbs.	Worsted and woollen	
	Berkshire	Southdown and black-faced Berkshire	Short	Fine	White	4½ lbs.	Worsted and woollen	
Sub-variety (e) Old Norfolk	Norfolk	Southdown and Norfolk or Downs	Short	Fine	White	3½ lbs.	Worsted and woollen	For flannels and low cloth.
Sub-variety (f) Old Wiltshire	Wiltshire	Southdown and Wiltshire	Short	Fine	White	3 lbs.	Worsted and woollen	
Sub-variety (g) Dorset	Neighbourhood of Dorchester	..	Short	Medium	White	3½ lbs.	Worsted and woollen	Livery cloth at Ilminster.
Sub-variety (h) Cornish	Cornwall	Cornish and Leicester	Long	Coarse	White	6 to 7 lbs.	Worsted and woollen	
Sub-variety (i) Old Lincoln	Lincolnshire	..	Long	Good	White	8 to 9 lbs.	Worsted	
	Lincolnshire Wolds	Lincoln and Leicester	Long	Good	White	7 lbs.	Worsted	
Sub-variety (j) Romney Marsh	Kent Southam Notts	Romney and Devon	Long	Medium		7 lbs.	Worsted	
Sub-variety (k) Bampton	Devonshire Buckland	Bampton and Leicester	Long	Very fine	White	8 lbs.	Worsted	
Sub-variety (l) Exmoor and Nott	Exmoor	Exmoor and Leicester	Long	Medium	White	4 lbs.	Worsted and woollen	
Sub-variety (m)	Devonshire	Cotswold and New Leicester	Long	Medium	White	7 to 8 lbs.	Worsted	
Sub-variety (n)	Dishley	..	Very long	Coarse	White	8 to 9 lbs.	Worsted	

Table showing the Varieties and Breeds of Sheep, Foreign and British—*Continued.*

Varieties and Sub-varieties.	Breed.	Cross.	Staple.	Quality.	General Colour.	Average Weight of Fleece washed.	Whether for Woollen or worsted.	General Application.
Sub-variety (o) Improved Tees-water	Durham York	Teeswater and New Leicester	Long	Fine		8 lbs.	Worsted	
Sub-variety (p) Woodland horned	Lancashire	Leicester and Woodland Southdown and Woodland				..		This breed is nearly, if not quite lost.
Sub-variety (q)	Lancashire	..	Long	Good	White	4½ lbs.	Worsted	
Sub-variety (r)	West Riding (Yorkshire)	Peniston and Leicester, Peniston and Cheviot	Short	Moderate	White	..	Woollen	
			Short		White	..	Woollen	
Sub-variety (s) Isle of Man	Manx Hills	..	Short	Fine	White and gray	2½ lbs.	Woollen	
	Manx Valleys	..	Long	Fine	..	7 lbs.	Worsted	
Sub-variety (t) The higher Welsh mountains	The mountain sheep	..	Short	Fine	White	2½ lbs.	Woollen	Chiefly used for flannels.
Sub-variety (u) Soft-woolled Welsh	The Anglesea		Medium	Not very fine	White	2½ to 5 lbs.	Worsted and woollen	
Sub-variety (c) Cannock Heath or Sutton Coldfield	Staffordshire		Fair length	Medium	White	6 to 7 lbs.	Worsted	Though generally much discoloured by smoke, it washes quite white.
Sub-variety (w) Cheviot	Northumberland		Medium	Medium	White		Worsted	
Sub-variety (x) Dun-faced	Northumberland, Scotland Westmoreland Cumberland							
Sub-variety (y) Black-faced	Northumberland Scotland		Medium	Coarse	White and gray		Worsted and woollen	

Sub-variety (z) Hebridean	The Hebrides	..	Long	Inferior	White		Worsted and woollen	
Sub-variety (aa) The Orkneys	The Orkneys		Long	Not very fine	White			
Sub-variety (bb) Shetland	Shetland	Shetland and Dutch	..	The finest	White	1¾ lbs.		
	The flounder-tailed		Long	Medium	White	4 lbs.		
Sub-variety (cc) Wicklow mountains	Cottagh	..	Short	Medium	White	2½ lbs.		Used for the flannel of Rathadrum, stuffs, etc.
	The Irish	..	Long	..	..	3 lbs.		
Sub-variety (dd) Herdwick	Cumberland hills		Short	Very coarse	White	3 lbs.	Woollen	Used only for low goods. This variety is remarkable for its hardiness and sagacity in foreseeing and preparing for a coming storm by the flock stationing itself in such a position as to cause the snow to drift, and thus leave a bare space for the sheep.
Sub-variety (ee) The Rass or Roosh	Bokhara							
3. Barwall sheep (Ovis Barual, Hodgson)	Nepal							
4. Hoonah sheep	Hoonah or blackfaced Tibet		Long	Soft and fine			Worsted	For ladies' dresses.
5. Cago (Ovis Cagio, Hodgson)	Cago or lame sheep of Cabul		Long	Fine				
6. Seling (Ovis Selingia, Hodgson)	Nepal, central hilly region and Eastern Tibet		Long	Fine	Some breeds black			For rugs and coverlets.
7. Curumbar	Mysore		Short }	Coarse	{ White Yellow Gray Brown Black }		Woollen and worsted	East Indian wools are chiefly used for making blankets, but small quantities are used also for making carpets and rugs — longest for worsted.
8. Gārār	India		Short }					
9. Dukhun	The Deccan		Short	Coarse				

Table showing the Varieties and Breeds of Sheep, Foreign and British—*Continued.*

Varieties and Sub-varieties.	Breed.	Cross.	Staple.	Quality.	General Colour.	Average Weight of Fleece washed.	Whether for Woollen or Worsted.	General Application.
10. West Indian	Jamaica		Short	Fine and soft, but mixed with hair			Woollen	
11. Brazilian	South America, Pernambuco							
12. Smooth - haired (Ovis Ethiopia, Charlet)	African		Not used					
13. African (Ovis Guinensis, Raü)	Senegal and Saharah		Not used					
14. Guinea sheep (Ovis Ammon Guinensis, Schreber)	The Guinea breed		Not used					
15. Moroant de la Chine	China		Short	Rather coarse, but peculiarly soft and silky	Yellow		Woollen	Blankets, rugs, and carpets.
16. Shaymbliar	India, Mysore		Not used					
17. Zeyla	Zeyla and Mokha		Short					
18. Fezzan	Tripoli and Tunis		Medium				Woollen	For making caps or fezzes.
19. Marocco (Ovis Aries Numimidæ, H. Smith)	Marocco		Short	Inferior, soft			Woollen	Used for felt goods, blankets, and rugs.
20. Congo sheep (Ovis Aries Congensis, H. Smith)	Congo		Not used					
21. Angola sheep (Ovis Angolensis, H. Smith)	Angola		Not used					

22. Yenu or Goit-ered sheep (Ovis Aries Steatiniora, H. Smith)	Angola		Short	Fine and close			Not used in Europe.
23. Ixalus (Ixalus probaton, Ogilby)							
24. Cretan sheep	Crete		Short and much curled	Soft and fine		Woollen	
25. Long - tailed (Ovis Longi-caudatus, Brisson);	Russia Odessa and Crimean	Russian and Merino	Long	Very soft	White		
	Wallachian		Long	Very silky, but mixed with hairs			
	Moldavian		Long	Superior, but mixed with hairs Fine	White		
	Greek Barbary		Short Hair not used				
	Donskoi		Medium	Coarse	White and gray	Woollen and worsted	
	Odessa	Merino	Short	Very fine	White	Woollen	
26. Broad - tailed sheep (Ovis lacticaudatus, Eraleben)							
Sub-variety (a) Fat-rumped sheep (Ovis stearopyga)	Tartarian, Indian, Syrian, Chinese, Russian, and South African		Long	Good			
Sub-variety (b) Persian	Persian		Long	Medium	White Black Fawn Yellow Brown Gray	Worsted	Used for musmuds; the unyeaned lamb skins for pelisses.
Sub-variety (c) Fat-tailed							

Table showing the Varieties and Breeds of Sheep, Foreign and British—*Continued.*

Varieties and Sub-varieties.	Breed.	Cross.	Staple.	Quality.	General Colour.	Average Weight of Fleece washed.	Whether for Woollen or Worsted.	General Application.
Sub-variety (d) Avra fiyel	Abyssinian							
Sub-variety (e) Bucharian	Bucharian, Caucasian, Persian, and Astracan		Short	Fine and much curled	Black, gray in unborn lambs			Much prized in un-yeaned state, when the delicate curled skins are dressed for furs.
Sub-variety (f) Tibetan	Tibetan		Long	Soft and fine				Used for dresses.
Sub-variety (g)	Cape of Good Hope		Fur-like, and used as such					As fur for trimming dresses, etc.
Sub-variety (h) Ovis Aries ap-pendiculata								
Sub-variety (i) Belkah	Palestine and Plains of Bel-kah		Short	Thick	White			
27. Many - horned sheep (Ovis polyceratus, Linnæus)	India and Ne-pal, Dumba							
28. The Pucha	Hindostan, Dumba							
29. Short-tailed	Northern Russia							
30. Sheep of Tartary	Tartary		..	..				None of these are found in our markets.
31. Madagascar	Madagascar		Short	Fine				
32. The bearded	West African		Hair not used	..				
Javanese	Java		Short and finely curled		White			

The foregoing table lays before us every known variety of breed of sheep, with many of the characteristics of their wools noted ; it only remains now for us to notice a few more points in relation to quality and other peculiarities.

Lord Western, the great amateur sheep breeder, in a letter to Mr. James Bischoff of Leeds, dated Felix Hall, February 1842, says : "I have always in the same house nine sheep bearing their fleeces for three years, three hoggetts, three shearlings, and three two shear ; when I take the third fleece I finish them for the Christmas slaughter ; I say finish, for they come very fat out of their fleeces, which weigh from twenty-five to thirty-two pounds ; these are kept for exhibition of the singular powers of the animal, both in production of such a fleece of wool and ability to support it, and of growing and thriving and fatting at the same time ; they are of course highly fed, but eat comparatively very little. I was induced to adopt this experiment from finding the fleece adhere as tenaciously to the skin at shearing time as at any other, which I thought was peculiar to the breed and their ability to carry their fleeces ; but I have recently seen a Leicester fleece of three years' growth of the extraordinary length of *three feet.*" This is indeed an extraordinary instance of length of growth, and shows to what an amazing extent the sheep yields itself to the culture and modification exercised by man. The student with a view to woollen manufacturing cannot too intimately acquaint himself with every peculiarity pertaining to wool and its nature, if he would acquire the knowledge and ability to use wool successfully in its various stages of manufacture, and such books as Dr. Bowman's *Structure of the Wool Fibre* cannot be too strongly commended to his attention, as nothing less than a thorough scientific and technical knowledge of both *material* and *processes* will enable him to face the fierce competition and pressure of the onward movement of our modern times. If the student does not take the trouble to inform himself of the nature and varieties of wool, he will never afterwards feel that *interest* in the material or take the *care* that he *otherwise would* in following it through the various processes of manufacture. A thoroughly grounded knowledge from the very beginning of the *nature and growth* of the raw material is the best foundation on which to build; *the perfection of the whole is the perfection of its parts individually,* in every minute particular from the very earliest point. How could we be induced to guard and *safeguard* the material upon which we are employed if we knew nothing of the *exceeding delicacy of its structure* and the *extreme sensitiveness* of its nature? The wool fibre is infinitely *more sensitive* and *much sooner injured* than any of the vegetable fibres, such as cotton or flax. It is as much more sensitive as the organisation of animal life is above that of vegetable life. We have seen in examining Dr. Bowman's plates on the wool fibre how wondrously the fibre is built up—layer upon layer of serrations, till their numbers per inch are counted by thousands,—how elaborately and exquisitely beautiful is the mechanism of its structure. We very inadequately realise in our own minds what is implied in the expressions "*six thousand* (6000) *scales or serrations per inch*," "planted upon a stem that is *only one three-thousandth* ($\frac{1}{3000}$) *of an inch in diameter*." We fail to realise in full the idea that such expressions imply, but we may perhaps realise it sufficiently to cease to wonder that such a delicately constructed fibre of animal production is so soon injured in the processes of manufacture. No other textile industry (only the worsted) has to deal with a raw material so sensitive and so soon injured as wool. In the course of manufacture there are three deadly enemies against which wool needs to be carefully, even jealously guarded; they are *excessive heat, strong alkalies*, and *strong acids;* we need to "red light" all these danger points on the road of its manufacture.

Referring to the manipulation of wool for worsted purposes one writer remarks, "The long wool of which stuffs and worsted goods are made is deprived of its felting properties. This is done by passing the wool through combs, which takes away the laminæ or feathery part of the wool and approximates it to the nature of silk." Another says, "Wool in the worsted manufacture is treated very much as cotton. The first process consists in abolishing those serrations of the wool which are of so much use in the making of cloth." In reference to the use of wool for woollen purposes one writer says, "They, the fibres, are torn and broken into innumerable minute fragments, and then mingled together in every direction." Another says, "It is on this account that the carding, which breaks it up, is so important." Important, forsooth, because it breaks up and damages the material! Can anything possibly be more misleading to the student and tend to lead him farther from the truth than talking of treating wool in this way? Surely we ought to have put forward a truer, sounder, and more intelligent teaching than this, especially when men with titles to their names from learned societies undertake to enlighten us on technical subjects pertaining to manufactures. Ought we not rather by all rational means to strive to retain the natural structure of the fibre, and to maintain its softness, strength, and elasticity uninjured? Of this we may rest assured, that it is no part of an accurately scientific mode of manufacture to do violence to the material upon which we are engaged, either by chemical or mechanical manipulation. We have neither right nor need to break the staple by any kind of rough mechanical usage, nor have we either the right or the need to subject the raw material to the action of destructive chemicals; ignorance only will allow either the one or the other; the intelligent manufacturer will allow neither.

In drawing our remarks on wool to a close we must not omit to mention again in a special manner one very striking manufacturing peculiarity that arises out of the construction of the wool fibre, viz. the great facility and capability for felting which the wool fibre possesses. All animal downs, hairs, or furs possess the power or capability of felting in a greater or lesser degree—even the hair of a dog will felt; but the wool fibres possess in an eminent degree this power or capability of felting whenever the fibres come into contact with each other in opposite directions.

Fig. 3 shows how the serrations interlock when the fibres approach each other, root end to root end, or the contrary. This great capability of felting in wool is a very important feature in the woollen manufacture, and contributes to the production of a great variety of cloths not otherwise producible, to great warmth and comfort in wear, and to a durability unsurpassed and unattainable by any other fabric.

Fig. 3 illustrates the position and tendency of bearing of the wool fibres towards each other when meeting in opposite directions, and it will be readily seen with what facility they interlock owing to the natural construction of the serrations surrounding the shaft of the fibre in such amazing numbers. The mouths of these serrations

open when under the influence of soap and a little
warmth, and when under these conditions wool fibres
are subjected to the pressure and friction of the fulling
machine they felt to an astonishing degree. This is a
distinguishing peculiarity, or property, which no vegetable fibre possesses, and which in combination with the
textural structure of the woollen fabric is of inestimable
value in the woollen manufacture.

PART III

OUR WOOL-SUPPLY

In reference to our general wool-supply, the first in
order is our own domestic or home-grown English wool.
Until recently great difficulty was experienced in ascertaining with any degree of accuracy the amount of
our home production. Since 1867, however, the Agricul-
tural Returns have been published, giving the number
of sheep and lambs on the 25th June in each year.
With these data to start from, and by estimating the
average weight of fleece per head, a more fairly reliable conclusion can be arrived at than was previously possible. The
number of sheep and lambs in the United Kingdom reached
34,837,597 in 1874; from them we derived a wool production of about 167,000,000 lbs., which was the highest on
record. The number of sheep and lambs in the United Kingdom on the 25th June 1891 was returned as 33,525,000, as
against 31,667,195 in 1890, showing further progress, being
the largest number since 1874. The clip was estimated at
148,000,000 lbs., as against 138,000,000 lbs. in 1890. These
figures are not absolutely reliable, as there is great reluctance

on the part of the farmers to furnish full returns of their live stock. Previous to 1780-90, and previous to the introduction of steam power into the woollen manufacture, our forefathers had to be content with a strong, firm, durable cloth made from home-grown wool, but their more polished and refined sons must be clothed in garments less rough to the touch, and more sightly to the eye, and this necessitated the introduction of Spanish wool, public taste calling for a still *firmer*, *closer*, and *compacter* fabric, that would carry a higher and more lustrous "finish." This demand still continuing, German wool began to be introduced about 1820, and superseded the Spanish, as it enabled the manufacturer to produce a closer - faced cloth. The German wool continued to be used till it was met by the cheaper merino wool of our own colonies, in the face of which, coupled with the rise of the fancy woollen manufacture, which admits of the use of a great variety of wools, its use gradually diminished, till in 1883 it had fallen to 16,138 bales, and in 1891 to 3335 bales. English wool is still, however, used freely in the production of army cloths, police clothing, upholstering purposes, and railway carriage cloths, besides shawls, blankets, flannels, fancy woollens, and worsteds; but even the Southdown, which is the finest class of English wool, has for some time been superseded in the manufacture of fine cloths.

The name of this finest class of English wool is derived from the South Downs, in Sussex, a range of hills some 60 miles in extent, abounding in short fine herbage, at a good elevation and with a dry healthy atmosphere. Large flocks of Southdown sheep are kept on these hills, the native breed being adapted to a dry hilly country, with bleak winds in winter and drought in summer. They are nimble, active animals, and as they thrive well on many other dry soils, they are often exported to the colonies and other countries for crossing purposes. They are small consumers of food, come early to maturity, with a ready aptitude to fatten, which renders them favourites with many from a mutton point of view, apart from any consideration as to the value of the fleece for manufacturing purposes.

Next in order of wool-supply, after our own home-grown wools, come the wool productions of our colonies. To the origination of this source of wool-supply the manufacturing public are greatly indebted to the intelligence, foresight, and perseverance of Mr. John M'Arthur, who at the beginning of the present century introduced the merino breed of sheep into the New South Wales portion of the Australian colonies. The example thus set was soon followed by Van Diemen's Land, and from these two, sheep were introduced into the sister colonies in succession, as they were settled, including New Zealand. The example of the Australian colonies stimulated the Cape to take action about 1826-27 and give attention to the growth of merino wool by crossing and improving the breed of their native sheep. Though our import of wool from our Australian and South African colonies commenced early in the first quarter of the century, still the quantity was so irregular and inconsiderable that our own public sales of these wools did not commence till 1839. In 1807 we received the first importation of merino wool from Australia, weighing only 245 lbs., and the next record is 167 lbs. for 1810, after which the supply became continuous; and while our import of German wool is gradually decreasing, our import of Australian and other colonial wools is just as steadily and constantly increasing, as will appear by comparing the figures of the following decades :—

IMPORTS OF WOOL FROM AUSTRALIA

1810	.	167 lbs.
1820	.	99,415 ,,
1830	.	1,967,309 ,,
1840	.	9,721,243 ,,
1850	.	39,018,221 ,,
1860	.	59,166,616 ,,
1870		175,081,427 ,.
1880		300,240,128 ,,
1890		418,702,000 ,,

Though our public sales of colonial wools did not commence till 1839, during which year there were catalogued for disposal 26,987 bales, the quantity has continued to increase with such wonderful rapidity that the product in 1887 amounted to 1,444,000 bales of the value of £20,216,000. The following particulars are copied from Messrs. H. Schwartze and Company's Annual Wool Report for 1891. The number of sheep in our Australian colonies according to official returns taken early in the year was as follows :—

	1892.	1891.	1890.	1889.	1888.	1887.
Queensland	20,289,633	18,007,234	14,470,095	13,444,005	12,926,158	9,690,445
New South Wales	61,831,416	55,986,431	50,106,768	46,503,469	46,965,152	39,169,304
Victoria	12,928,146	12,736,143	10,882,231	10,818,575	10,613,959	10,700,403
South Australia	7,646,239	7,004,642	6,432,201	7,150,000[1]	7,254,000	6,696,406[2]
Western ,,	1,962,212	2,524,903	2,366,681	2,112,393	1,909,940	1,809,071
Tasmania	1,664,118	1,619,256	1,551,429	1,430,065	1,547,242	1,608,946
Total Australia	106,321,764	97,878,609	85,809,405	81,458,507	81,216,451	69,674,575
New Zealand	18,128,186	16,749,692	16,116,114	16,677,415[3]	16,677,445	16,564,595
Total Australasia	121,449,950	114,628,301	101,925,519	98,135,952	97,893,896	86,239,170

[1] Estimated.   [2] Returns for 1885.   [3] Returns for 1888.

The total of 114,000,000 compares with 65,000,000 in 1880, 50,000,000 in 1870, and 20,000,000 in 1860. The latest return gives 124,500,000 for 1892.

The proportion of grease, fleece, and scoured, and the quantity of cross-bred, catalogued in London are shown in the following :—

	1891.	1890.	1889.	1883.	1876.	1869.
	Bales.	Bales.	Bales.	Bales.	Bales.	Bales.
Grease	990,000	744,000	794,000	679,000	387,000	158,000
Fleece	24,000	34,000	25,000	101,000	238,000	266,000
Scoured	338,000	305,000	292,000	273,000	157,000	114,000
Total catalogued	1,352,000	1,083,000	1,111,000	1,053,000	782,000	538,000
Of which Australian Cross-bred	58,000	51,000	54,000	92,000	49,000	7500
New Zealand Cross-bred	215,000	185,000	182,000	117,000	53,000	7500
Total Cross-bred catalogued	273,000	236,000	236,000	209,000	102,000	15,000

The quantity of fleece washed is now less than 2 per cent of the total, and comprises only 10,000 bales Port Phillip and 5300 bales Sydney. Queensland fleeces have disappeared from the market. New Zealand cross-breds now form a large percentage of the total production of the colony; they increased by 30,000 bales last year, while of merino wools there were 10,000 bales less. These figures indicate that New Zealand cross-breds are becoming quite favourite wools in the market, and there are good and substantial reasons for such a preference. Like all wools grown on small islands, they can always be relied upon for being sound, soft, and kindly in the hand, and the explanation for this is that even in seasons of the greatest drought there is always a sufficiency of moisture in the atmosphere that the animal can take in with its breath, so that its wool never becomes tender in its growth, even during the greatest drought, as the distance from the sea can never be great in any part of the country. Such wools are invariably good, milling quickly and well. Though Australia itself is an island, yet compared with New Zealand and Tasmania it must be considered a continent, and wools grown in the inland districts at a great distance from the sea will always need to be carefully examined by the purchaser after a season of drought, as they will always be liable to have tender places in the staple that very greatly reduce the value.

The import of Cape wools into England for the calendar year is 316,510 bales, against 283,494 bales in 1890; into the continent direct, 6324 bales, against 2663 bales in 1890. The following gives the production of South Africa during the last four seasons :—

	Import Season.			
	1891. 26th Nov. 1890. 24th Nov. 1891.	1890. 27th Nov. 1889. 25th Nov. 1890.	1889. 28th Nov. 1888. 26th Nov. 1889.	1888. 23rd Nov. 1887. 27th Nov. 1888.
Cape—	Bales.	Bales.	Bales.	Bales.
Into England (for the five series) .	307,694	278,719	302,816	280,186
(Of which for London market) .	152,000	177,000	172,000	168,000
(,, ,, forwarded to Interior) .	58,000	22,000	30,000	21,000
(,, ,, ,, ,, Continent)	95,000	79,000	100,000	91,000
(,, ,, ,, ,, America) .	3000	1000	1000	...
Into Continent direct . . .	6324	2663	...	1398
,, America direct . . .	7686	6724	7103	7968
Total . . .	321,704	288,106	309,919	289,552
Viz. Western Province .	26,933	23,482	28,231	24,190
,, Eastern ,, . .	217,390	190,037	203,223	189,418
,, Natal and Interior .	77,381	74,587	78,465	79,944

There was an increase of 34,000 bales, almost entirely in the Eastern Province. The clip, on the whole, was better than the previous one, although there was a drought in some parts at the commencement of the season in 1890.

The proportion of grease, fleece, and scoured catalogued in London is shown in the following :—

	1891.	1890.	1889.	1883.	1876.	1869.
	Bales.	Bales.	Bales.	Bales.	Bales.	Bales.
Grease . .	100,000	118,100	106,200	43,300	16,600	12,800
Fleece . .	11,500	15,500	15,000	19,000	43,500	58,000
Scoured . .	49,400	56,000	50,800	84,000	82,800	69,500
Total catalogued	160,900	189,600	172,000	146,300	142,900	140,800

The direct purchases, particularly on the part of the home trade, have been considerably larger than in previous years, and only 156,000 bales, or a little less than half the clip, were sold in London, against 173,000, 172,000, and 170,000 bales in the three preceding years.

The following gives the yearly total value since 1885 of the colonial supply, based upon a fairly trustworthy average value per bale :—

IMPORTS INTO EUROPE AND AMERICA FOR THE SEASON

Year.	Australian Bales.	Cape Bales.	Total Colonial Bales.	Average Value per Bale.	Total Value.
1885	1,094,000	188,000	1,282,000	£14	£17,948,000
1886	1,196,000	236,000	1,432,000	$13\frac{1}{2}$	19,332,000
1887	1,207,000	237,000	1,444,000	14	20,216,000
1888	1,315,000	289,000	1,604,000	$13\frac{1}{2}$	21,654,000
1889	1,385,000	310,000	1,695,000	$15\frac{1}{2}$	26,272,000
1890	1,411,000	288,000	1,699,000	$14\frac{3}{4}$	25,060,000
1891	1,683,000	322,000	2,005,000	$13\frac{1}{2}$	27,067,000
1892	1,835,000	291,000	2,126,000	12	25,512,000

The increase of 306,000 bales, or 18 per cent, in the colonial supplies is faced by a fall in the average price or value per bale of £1¼, or about 8 per cent, and the total value of the clip in 1891 was larger by two million pounds sterling than in 1890. The wool clip of the world is said to have increased fivefold in the fifty years intervening between 1832 and 1882.

In 1891 it is calculated that the United Kingdom produced 148,000,000 lbs., the Continent 450,000,000 lbs., North America 316,000,000 lbs., while 592,000,000 lbs. were imported from Australia, 102,000,000 lbs. from the Cape, 330,000,000 lbs. from the River Plate, and 179,000,000 lbs. of sundry "other sorts," making a total of 2,117,000,000 lbs. This large amount was distributed in the following manner amongst consumers :—

United Kingdom .	487,000,000 lbs.
The Continent	1,174,000,000 lbs.
North America	456,000,000 lbs.

Taking the population of Europe and North America at 414,000,000, this shows an average consumption for the year of 5·11 lbs. per head. The totals show 8 per cent higher than the figures for 1890, but only 5 per cent over 1889.

For many years to come, says Consul Baker, of Buenos Ayres, speaking of River Plate wools, the leading industry of the Argentine Republic must continue to be the raising of sheep and the production of wool. The history of Argentine sheep farming dates back to 1550, when the merinoes were introduced from Spain; but it received no attention whatever from the early settlers of the country. Even as late as 1840 sheep were of little value from a commercial point of view. They were allowed the run of the Pampas as a cheap but not a desirable article of food. The wool was not considered to be worth the expense of carting it to the town, and was often used for litter and other like purposes.

The first great impulse to the sheep farming industry was given about thirty-seven years ago, when a few Scotch and Irish sheep farmers, seeing the wonderful possibilities which the Argentine Republic possessed for producing sheep, began to improve the stock by the importation of the finer Negretti and Rambouillet breed. In 1852 the number of sheep in the entire Republic was 5,500,000, but with the increased fineness and general improvement of the wool a foreign market opened out and a demand sprang up, and in 1860 the number of sheep had increased to 14,000,000. Then followed the American War, which caused an unprecedented demand for foreign wools, and sent the prices up to fabulous figures. Everybody in Argentina who could buy, or in any way raise money to get hold of a flock of sheep, went into the business, and in 1867 the number of sheep was estimated to be 40,000,000. In 1888 the official statistics gave the number as 66,701,097. The number of sheep in the Argentine Republic is at present equal to 18,000 per 1000 inhabitants, which, Consul Baker believes, is in excess of that of any other country in the world in proportion to the population. The wool clip of the country bears a very small proportion to the number of sheep. The yield is only about 3·8 lbs. of "dirty" or unwashed wool, and by being shipped in the grease, River Plate wool suffers more from shrinkage than either Australian or Cape wool. There are no reliable figures available for accurately ascertaining the value of the Argentine wool clip.

Consul Baker, who is one of the best authorities on all matters relating to the Argentine Republic, says there is no doubt that sheep-growing, along with the cattle business, must continue to be one of the leading industries of the Argentine Republic. Its development can be indefinitely extended. There are millions of acres all through the vast interior, and running along the Andes away down to the extreme limits of Patagonia, which, whatever else may be said of the country, can be used for sheep and cattle pasturage; and as the inside lands become too valuable for sheep and cattle ranges, and are employed for farming purposes, these outside "camps" will come into requisition for pasturage. Already, with the progress of agriculture, the sheep and cattle are moving farther and farther out, and with the development of the country we may ultimately expect to see all the vast extent of Pampa, now beyond reach of the railways or markets, filled with the sheep and cattle estancias. Whatever may be the fate of wool and wool-growing in other countries, it may be said to be safe in the Argentine Republic for all time to come, and this will undoubtedly be one of the great sources of wool-supply in the future.

Manufacturers in the States have found the cost of the raw material so enhanced that in many classes of goods they are still undersold by English and continental competitors. The demand for domestic wools has declined because the foreign wools used with them cannot be profitably imported. The result is that while for some years prior to the M'Kinley tariff coming into operation wool-growing in the United States had been a progressive industry, last year it underwent a serious change in the opposite direction, not from any adverse causes such as are known to pastoralists in New South Wales, where droughts, rabbits, and various other difficulties have to be contended with. This will be more clearly seen and understood by the following returns, compiled by an undoubted authority :—

United States Wool Clip.	lbs.
1888 .	309,000,000
1889 .	302,000,000
1890	316,000,000
1891 .	285,000,000

With regard to the future of wool, the outlook may be said to be hopeful rather than otherwise, all the available evidence pointing in the direction of moderate expansion rather than contraction of supply.

Importation of Colonial and Foreign Wool into the United Kingdom from 1882 to 1891.

	1891.	1890.	1889.	1888.	1887.	1886.	1885.	1884.	1883.	1882.
New South Wales	353,407	274,448	306,091	321,154	245,290	265,181	217,119	241,277	234,659	230,284
Queensland	173,558	144,093	126,637	122,867	106,614	84,065	104,361	99,974	60,858	54,093
Victoria	365,490	365,172	372,057	380,336	345,396	360,731	317,152	358,228	336,518	364,041
South Australia	120,665	98,249	111,236	115,849	106,403	130,628	115,108	118,357	108,487	122,167
Tasmania	25,855	23,537	22,035	20,167	22,261	21,463	21,681	24,415	24,038	23,429
Western Australia	26,933	27,949	22,897	19,382	17,656	16,862	14,427	13,204	11,208	11,615
New Zealand	315,055	292,724	277,726	265,684	272,918	260,912	237,875	228,900	215,024	194,102
Australian	1,380,963	1,226,172	1,238,679	1,245,433	1,116,538	1,139,842	1,027,723	1,084,355	990,792	999,731
Cape	316,510	283,494	287,334	288,910	234,728	227,289	182,168	189,377	187,368	191,113
Colonial	1,697,473	1,509,666	1,526,013	1,534,343	1,351,266	1,367,131	1,209,891	1,273,732	1,178,160	1,190,844
German	3,335	4,592	7,358	5,356	9,589	12,005	9,700	10,045	16,138	10,297
Portuguese and Spanish	8,941	13,538	21,558	18,157	16,385	24,355	7,731	7,598	10,683	10,849
East India	104,310	103,342	113,518	105,240	100,646	101,770	83,595	75,061	78,708	83,562
Persian	29,457	22,328	27,350	28,930	23,299	16,755	10,104	15,680	7,486	8,444
Russian	96,205	65,506	106,263	59,802	66,422	65,027	63,368	48,635	56,212	44,201
River Plate	9,145	6,310	11,885	10,350	7,016	12,440	8,728	4,305	5,690	5,868
Peru, Lima, and Chili	30,328	30,059	32,410	27,096	35,576	23,209	31,270	54,336	10,690	39,466
Alpaca	31,740	27,441	34,637	29,139	34,366	26,718	34,421	67,133	13,576	38,872
Mediterranean and African	64,795	74,237	88,395	78,980	88,371	62,678	33,127	20,874	36,799	36,957
Mohair	62,993	48,131	77,526	72,767	56,005	76,690	52,547	66,873	55,895	66,202
Sundry	47,433	36,736	44,981	32,800	24,369	30,404	25,325	15,333	18,423	17,809
Total bales	2,186,155	1,941,886	2,091,894	2,002,960	1,813,310	1,819,182	1,569,717	1,659,603	1,488,460	1,553,371

The following table and paragraph are taken from "Wool and its Uses" in *The Textile Manufacturer* for June 1888 :—

Comparison of the Quantity of Foreign and Colonial Wool of all kinds imported into this country, and exports of the same, with the production of English wool, during twenty-one years, in round figures : [1]—

Year.	Imports of Foreign and Colonial Wool.	Exports of Foreign and Colonial Wool.	Leaving of Foreign and Colonial Wool for Home Consumption.	Home Production.	Year.
	lbs.	lbs.	lbs.	lbs.	
1867	236,000,000	91,000.000	145,000,000	157,000,000	1867
1868	260,000,000	105,000,000	155,000,000	166,000,000	1868
1869	263,000,000	117,000,000	146,000,000	156,000,000	1869
1870	266,000,000	93,000,000	173,000,000	150,000,000	1870
1871	332,000,000	135,000,000	198,000,000	145,000,000	1871
1872	313,000,000	138,000,000	175,000,000	156,000,000	1872
1873	325,000,000	123,000,000	202,000,000	165,000,000	1873
1874	352,000,000	144,000,000	208,000,000	167,000,000	1874
1875	372,000,000	172,000,000	200,000,000	162,000,000	1875
1876	396,000,000	173,000,000	223,000,000	156,000,000	1876
1877	418,000,000	187,000,000	231,000,000	152,000,000	1877
1878	407,000,000	199,000,000	208,000,000	152,000,000	1878
1879	427,000,000	243,000,000	184,000,000	153,000,000	1879
1880	475,000,000	237,000,000	238,000,000	149,000,000	1880
1881	460,000,000	265,000,000	195,000,000	139,000,000	1881
1882	505,000,000	263,000,000	242,000,000	129,000,000	1882
1883	509,000,000	277,000,000	232,000,000	128,000,000	1883
1884	544,000,000	277,000,000	267,000,000	132,000,000	1884
1885	520,000,000	268,000,000	252,000,000	136,000,000	1885
1886	615,000,000	310,000,000	305,000,000	136,000,000	1886
1887	597,000,000	319,000,000	278,000,000	134,000,000	1887

[1] 1867 was the first year when reliable statistics of the growth of English wool were available.

The influence of foreign and colonial wools on the price of English wool is enormous. Twenty-nine years ago the value of the English clip was roughly estimated at £13,000,000, but in 1888 it was estimated at £6,000,000. A comparatively small proportion of the serious loss to the farming interest is due to the falling off in quantity. It is interesting to note in connection with these figures that while in 1866 the value of a bale of colonial wool was about £24, the total value of colonial wool imported was £11,735,000, whereas in 1887, with a value per bale of about £14, the total value was £20,216,000. After allowing for the fact that a larger proportion of the wool is now imported in the grease than was the case twenty-nine years ago, the figures are sufficiently startling. The colonial wool imported in 1887, if taken at the prices of 1866, would show a value of £34,656,000. The price of alpaca in 1866 was 3s. 4d. per lb. ; in 1886 it was about 1s. per lb. The value of mohair in 1866 was 3s. 8d. per lb. ; in 1886 it was 1s. 2d. per lb. There are many points about the manner in which colonial wool is sent to market and dealt with which give it an enormous advantage over our own. The flocks are often very large, and after being shorn the wool is generally carefully and thoroughly skirted, *i.e.* the short wool growing round the neck and legs and down the belly of the animal is taken off, and packed into separate bales. The wool is also classed into different descriptions, merino and cross-bred not being mixed in the same bales, except in some of the smaller flocks. The consequence is that on its arrival in London large quantities can be taken direct to the comber without any sorting whatever, and the same thing is often done in the fancy woollen trade. A spinner or manufacturer can go round the warehouses and select the exact sort he wants ; and sometimes during one evening he can buy his lots and have done with this branch of his business for some time. As the London wool sales generally last from three to six weeks, and as there are seldom less than 10,000 bales offered every night, there is plenty of choice.

When this style of business is compared with the dilatory and unbusiness-like manner of buying English wool from the farmer, it will be seen what an immense saving of time and trouble there is to the user of colonial wool as compared with the user of English. A manufacturer can, and often does, purchase as much wool in London in a single night as would take him a month to buy in Lincolnshire or Shropshire. Very few people, except those having actual experience, have any idea of the vast variety of wool which is to be bought at the London sales. Almost every English sort can be matched there, and where the object to be aimed at is fineness of texture and softness of handle, the London colonial sale is the market to go to; there the finest-grown colonial merinoes are to be met with; at least 75 per cent of all the merino wool produced is used in the woollen manufacture.

In conclusion, it may be remarked in the interests of the wool industries that it is very desirable that more of the wool should reach this country in the grease, that there should be less scouring of wool in the colonies and elsewhere, that the user of the wool, the manufacturer, should be the only party to scour the wool, and that every due regard should always be paid to retaining the natural properties of softness, strength, and lustre by avoiding all use of strong fixed alkalies in scouring and high heat in the drying.

PART IV

WOOL SORTING

AFTER the purchase of the wool the first operation in the long series of processes in the woollen manufacture is the sorting of the wool. Perhaps there is no operation or process in a modern mill that varies so much as that of sorting the wool. When only plain, superfine, highly-faced goods were produced, great care and nicety of judgment had to be exercised ; but with a lessened production of such goods, and the rise of the fancy woollen manufacture, with its almost endless variety of pattern and makes of cloth, the sorting of the wool is now very far removed from being a " fixed quantity," as each district has some peculiarity to work up to and cater for in its prevailing style of cloth, which again changes, more or less, with each of the season's patterns, as they follow each other in rapid succession of form and colour, like the changes of the kaleidoscope. Formerly the wool stapler sorted the fleeces, as soon as convenient after purchasing them, into the respective varieties of staples or locks of equal quality of which his fleeces were capable, and according to the varying demand of his customers, and kept his wool in this assorted state, ready to supply the market demand ; one manufacturer wanting a fine quality staple, another a medium, and a third a low quality staple, each found at the wool stapler's a supply of such kinds and qualities as suited their respective wants and purposes. Wool was thus purchased wholesale by the staplers, then sorted and retailed to the small manufacturers in small quantities at a time, say sufficient for a " piece-wool"or more frequently " a couple of piece-wools," which was carried on horseback over the moors to the distant homes of the small manufacturers, frequently on the return journey from market after disposing of the " piece " or two of manufactured cloth ; and hence we find that the dealers in wool were called wool staplers, because they kept wool for sale, ready assorted into the respective staples or qualities for supplying the small manufacturer, who frequently purchased no more than a couple of "piece-wools" at a time.

Since then "the times have changed," and the wool staplers of the present day are no longer wool staplers in reality, but have become wool merchants, buying and selling wool only in a wholesale manner; some few sort a little, but on a very different scale to the wool staplers of bygone days. The wool sorter, with the march of time, has had to leave the wool merchant's warehouse and to betake himself to the mill, and there he plies his vocation in separating the different portions of the fleeces of wool into various classes, qualities, and lengths of staple, and other characteristically different sorts, to suit the requirements of the particular kind of manufacture upon which the mill for the time being may be engaged; and thus the wool sorter performs the first of the series of those complicated and intricate operations that constitute the routine and characteristics of a modern woollen manufactory. The spinner or manufacturer in selecting and purchasing wool selects such as will yield the greatest bulk of the quality he is in immediate want of, and which will leave the smallest number of sorts and weight in stock for which he has no immediate use. Sometimes there are only three sorts made, a "first," "second," and "third," with perhaps a little very coarse "britch"; in other cases only "first" and "second," with "britch"; and in many cases for fancy woollens only the coarse skirt ends are thrown off. There is now no absolute rule or standard; the exigencies of the manufacture determine the mode to be followed, and are the only guiding rule.

A single fleece of wool is capable of yielding a great variety of qualities, if sorted to the utmost nicety, as will be observed by noting the markings on the following figure :—

The forepart of the animal yields the finest wool, particularly the shoulder, marked 1. That portion marked 2 is next in fineness. Then 3 on the neck side of the shoulder follows, and so on to the end of the markings; only it must be understood that there is no hard and fast dividing line, the qualities running into each other as they approach one another on every side. The practised eye of the wool sorter follows the quality to the boundary markings or beyond, just according to the number of qualities

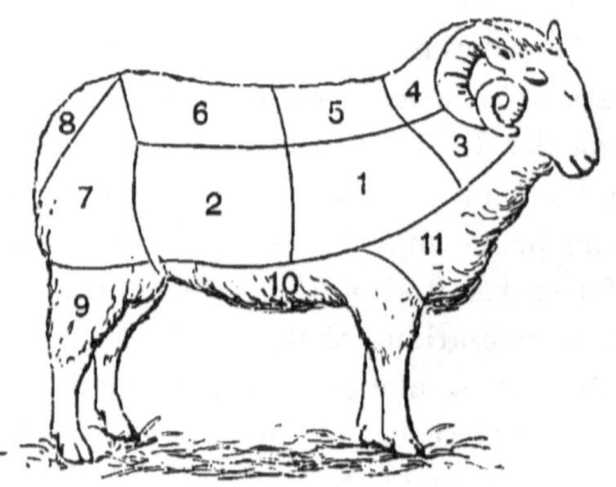

Fig. 7.—Qualities of Wool.

into which he is sorting the fleeces before him for the time being, because at one time he may be sorting the same pile of fleeces into twice as many qualities as at another time, according to the requirements of the firm's manufacture. There is no operation connected with the woollen manufacture that has varied more than that of wool-sorting during the last hundred years. It has always had to accommodate itself to the requirements of the manufacturer, and during that period those requirements have varied to an extreme degree.

PART V

WOOL SCOURING AND DRYING

THE proper scouring of wool is a much more important operation than it has hitherto been considered in the general management of our woollen manufactories. Before the introduction of wool-scouring machines it was a very laborious piece of work, and the man that was strongest in the back, though he was weakest in the head, was usually selected and sent to the wool-scouring corner, whereas the scouring of wool as it ought to be scoured requires the greatest nicety of judgment, intelligent care, and skilful attention. The value of the material alone that passes through the hands of the wool scourer in a single day in one of our large manufacturing works amounts to a very considerable sum; but this is not all; considering how much all the after processes depend upon the right scouring of the wool, the subject has not so far received in this country a tithe of the attention that its merits demand. On the continent the subject, for hundreds of years, has received much more attention than it has been deemed worthy of in England. So long ago as 1723, in Prussia, according to a footnote on page 23 in *Southy on Colonial Wools*, it was "ordained that the approved method of washing or scouring and sorting wool should be read from the pulpit," so important was it deemed to have these early processes rightly attended to. Without the proper washing and cleaning of the wool it is impossible for the dyer to get brilliant and fast colours; without proper scouring of the wool it is impossible for the carding engineer to card it properly, as the wool will not open freely on the carding machines if it be imperfectly scoured. If the wool cannot be made to open freely by the carding engineer, but breaks up instead of opening, it is impossible for the spinner to get a sound and even thread out of it, and if the spinner cannot produce a sound, clear, even thread, the weaver cannot produce a perfect piece of cloth, his work being spoiled by being full of damages and imperfections through the breakage of the unsound yarn. It may easily be understood how very much the processes depend one upon another for being able to reach the known standards of practical perfection pertaining to each, and in no instance have we this more strikingly illustrated than in connection with the scouring of the wool. It is one of the earliest of manufacturing operations, and if imperfectly performed affects very seriously the well-being of all the operations that follow. The scouring of wool is the cleansing of it from its yolk, natural grease, and dirt, preparatory to its being manufactured, and this should be set about from one standpoint, *that the work must be accomplished by the mildest means possible*, so as to retain as much as possible of the natural softness, strength, free openness, and brilliancy of the fibre. If we destroy the natural softness of the fibre by any kind of harsh treatment, we at the same time destroy its soundness; for as in growth, so in the manufacture, lose the softness and you lose the soundness. In an exact ratio these two leading features go hand in hand, and cannot possibly be separated; present or absent, they are ever in close company, and no effort is too great to retain their presence through every process of the woollen manufacture to the end. And with the view of accomplishing so desirable a consummation as this, much more can be done at the very outset of the

manufacture by rightly scouring the wool than has hitherto been considered attainable.   A great portion of the yolk
is soluble in water; otherwise how can the sheep be cleaned
by simple immersion in water preparatory to shearing?
English sheep used always to be washed immediately
before shearing, and in every sheep-farming neighbourhood
there were "wash-pits" in the nearest brook; the name
still remains attached to certain localities, though the sheep
and sheep-washing have both disappeared in consequence
of the springing up of towns.   The yolk of the sheep's
wool is a sort of natural soap, having a basis of potash,
and owes its odour to a small quantity of animal oil,
which, along with the potash, constitutes it a true natural
soap — hence its solubility in water — and this would
admit of the wool being almost thoroughly and completely washed in a stream of running water were it not
for the presence in the fleece of a portion of fatty matter,
uncombined, chemically, with the yolk, which remains
attached to the wool fibres, and refuses to pass off by the
use of water only.   But so far as the cleansing of wool
can be carried on by water alone, so far ought we to carry
it, and having got so much dirty matter out of the way,
we ought to resort to alkaline solutions, when all that
is required can be done with a much weaker solution than
would otherwise have been possible.   It is admitted on
all hands that great care is necessary when we use sharp
and dangerous instruments, and equal care and consideration ought to be used when we apply alkalies to wool, for
this good and sufficient reason, that wool, subjected to the
action of strong alkalies, dissolves and is entirely destroyed,
so that in subjecting it to the influence and action of
alkalies we are, as it were, "blood-letting," we are taking
away so much of the life element of the wool.   Heat

carried to excess acts upon wool in like manner; for when, in scouring, soda ash or any allied soda compounds are used, and the heat employed exceeds 120° F., the soda acts upon the wool very adversely, as wool is very sensitive to the action of alkalies, *especially* at high temperatures. Superheated steam will decompose wool and reduce it to a perfect jelly; even pure water heated to 230° F. will decompose it, so that when subjecting wool to the boiling-point we are only 18° of heat from the point of its destruction. If, then, wool, being subjected to an excess of heat and alkali, begins to decompose, is it not therefore clear that every precaution should be taken to defend it from harm, even from the very outset of its manufacture? If, therefore, in the scouring of wool we are treading on such fatally dangerous ground, seeing that we cannot even clean it without going *a considerable way on the road towards its destruction*, let us take heed to our steps and do the least possible mischief; let us do everything we can that will enable us to *keep the heat down* while the wool is *under the action of alkalies*, and let us also do everything that we can to remove as much of the yolk and dirt from the wool by the use of water alone before we begin to use alkalies, for if we can get rid of a quantity of the yolk and dirt in this way, by the use of water alone at low temperature, it will be quite clear to an ordinary comprehension that we can accomplish the remaining portion of our work by a much weaker scouring liquid than if we had attacked the yolk and the dirt in gross bulk. An apparatus was shown in the Crystal Palace Exhibition of 1881, by Mr. Scharr of Bradford, for forcing water through the wool preparatory to scouring; but whether such an apparatus as Mr. Scharr's or a simple steeping of the wool and passing it through rollers is to be resorted to, I will not presume to dictate, but must content myself by merely indicating and suggesting a course of action, as each manufacturer will consult his own convenience. But the importance of reducing the severity of the operation of wool scouring, both chemically and mechanically, cannot be too strongly urged upon the attention of the manufacturer. Formerly wool used to be scoured by steeping it for a time in a liquid composed of two parts water to one part of old urine, heated to a point at which the operator could just bear to dip his hand into the liquor. The urine used to be collected by going from house to house, and was stored in vessels in order that it might undergo the changes by which the ammonia would be developed, being much more effective when it had become " old," as it was termed.

The scourer used to take a portion of wool and soak it for a time in a pan about two-thirds full of liquor composed as above, gently stirring the wool occasionally, and then at last forking it out and placing it on a rack to drain over the pan, after which he would convey it to a cistern of water and wash it, and get rid of the remaining sud and dirt. The liquor now used for wool scouring is not of the same mild inoffensive kind as formerly ; we have changed, but changes are not always improvements, and certainly they have not been so in the use of wool-scouring materials. A great portion of the urine used of old was ammonia, and as such was volatile and passed off, leaving the wool kind and soft to the touch, and very rarely injured ; but now cheaper agents are resorted to in the shape of fixed alkalies, soda in some of its forms, such as soda ash, being most frequently the article employed. Now when a fixed alkali is resorted to, it does not at once appeal to the sense of smell, if used in excess, nor does it leave the wool to go on

its way in peace, as a volatile alkali would, but it attacks
the very vitals of the wool, and never fails to gnaw at
those vitals every time the material or cloth is wetted
afresh, doing battle against the best efforts of the dyer, the
carder, and the finisher, through all the succeeding processes. *Wool ought never under any circumstances to be scoured
with a fixed alkali, but always with a volatile one.* Why not
return to the mild action of urine again as the scouring
agent for wool? As much urine could easily be collected
about a manufactory by means of a little piping as would
scour all the wool used there, instead of being allowed to
run to waste down the drains, and the manufacturer buying
so fearfully injurious a scouring agent for his wool as soda,
in any of its various forms, or any other fixed alkali. The
manufacturer cannot have too strongly impressed upon his
attention the necessity of using only a mild *volatile alkali*
for wool scouring and *a low heat*. Since the days of clear
river-washed wool in this country there have been several
methods to clean and dry wool before it is combed or
carded. When the rivers and brooks became dirty, owing
to increased manufacture and population, the wash-pits
had to be abandoned, and the wool hand-washed in cisterns
of warm water assisted by a ley or washing liquor composed of urine and water, as intimated above. As the
demand for large quantities of clean wool increased, wool-
scouring and washing machines were introduced into the
large factories as labour and water-saving machines. Many
designs of these machines are still in use, but for the last
few years, say since the Bradford Technical School Exhibition of 1882, there have been *two* distinct and recognised
methods of scouring and washing wool by machinery.
Briefly, it may be said, one method is agitation in the
scour liquor and then squeezing; the other method is

steeping with the least possible agitation and then squeezing.

The machine used in the first method is that which is known as the swing rake machine, by Petrie, in which the wool is passed through a long cistern or bowl, fitted up with swing rakes (see Fig. 8), which agitate and pass the wool quickly through the washing liquor up to the lifting apparatus, and so further forward up to the squeezing rollers. The agitation given by the swing rakes assists in loosening and getting out the sand and dirt and other extraneous matter whilst the wool is passing through the scour or washing liquor. This is a very useful machine for wool scouring when the wool put through is not liable to be injured by the agitation in the scouring

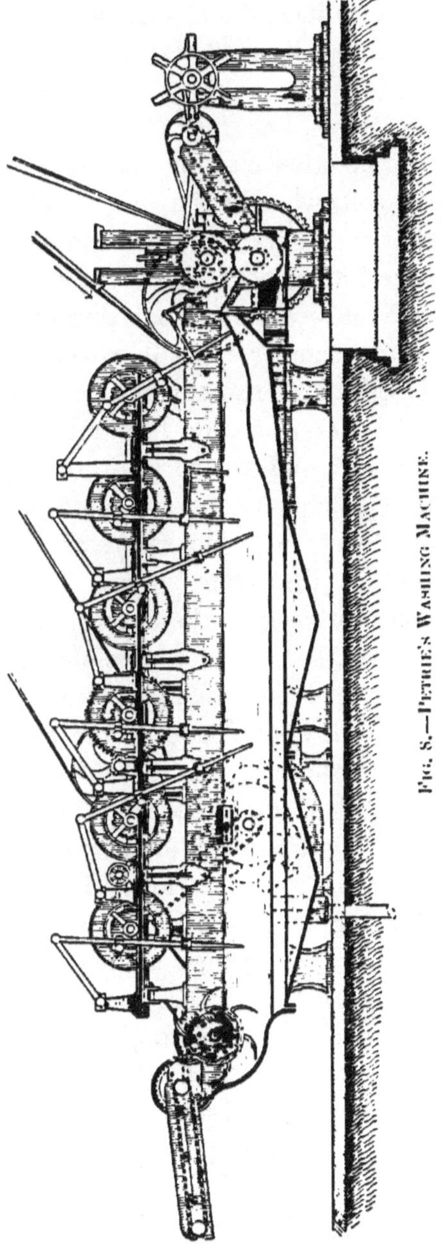

Fig. 8.—Petrie's Washing Machine.

liquor; and even in such cases as this the difficulty can be overcome by moderating the speed of the rakes, but for cleaning some kinds of short material the agitation is preferable.

The other class of machines for steeping and squeezing are somewhat similar in general outline, but so constructed that the rakes, or rather the series of rakes, formed into a rake frame move very slowly through the washing liquor, and thus pass the wool through in the most regular and gentle manner in bulk without agitation or disturbance. This class of machine answers very well for very fine long-stapled merinoes or for long fine Botany wools generally, and for long-stapled wools of any kind that are intended for combing, as there is no tendency to mat or felt, or in any way to tether the wool by agitation in passing it through the scour liquor. Fig. 9 shows the rake frame down at work in the cistern, moving the wool forwards in the gentlest manner possible until the frame reaches the end of the cistern close to the squeezing rollers, when it lifts bodily, by means of the chains and pulleys, clear out of the liquor and travels back to the forepart of the cistern to commence the journey afresh and carry the wool another stage farther along the cistern towards the squeezing rollers. This action is repeated until the wool is thus carried forward to its destination.

The most general method of using these machines is to have a "set" of three machines following each other in continuous line, the wool dropping from one to another. The first machine cistern contains the scouring liquor, and the second machine cistern contains a semi-scour, about one-third the alkaline strength of the first cistern and about half its heat. The practice too commonly followed of dropping the wool directly into cold water immediately after leaving the

squeezing rollers from the hot scour of the first machine
is too sudden a change, and chills the wool and sets
any remains of grease that may still be left in it. By
adopting a middle cistern all this is obviated; the
remains of grease get loosened in going through a second
cistern, and after passing through the second set of
squeezing rollers the wool may be let fall into the
cold water of the third cistern and get a thorough wash.
This last cistern contains nothing but cold water, and
should be supplied by $1\frac{1}{2}$ to 2-inch pipe kept continu-

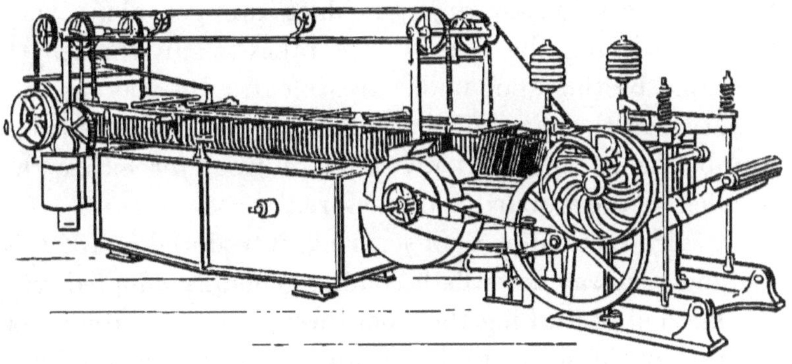

Fig. 9.—Steeping and Squeezing Machine.

ously running all the time the machine is at work, so as
to constantly renew the water. An overflow should be
placed at the opposite end of the cistern to that of the
inlet pipe, to allow the water to escape as it becomes
dirty.

The heat of the scour in the first cistern should on no
account be allowed to exceed 120° F. This is a point
that cannot be too carefully guarded, for wool, while under
the influence of alkalies, as during scouring, is very
injuriously affected by great heat.

Each machine has a set of squeezing rollers at the

delivery end, and outside the delivery end of the last machine of the series there is a small wince (see Fig. 8), which revolves quickly and beats open the wool as it emerges from the squeezing rollers. In small manufactories where only one set of scouring machines is in use the swing-rake kind is generally selected, and by accommodating the speed of the rakes to the kind of work in hand, the material need not be tethered or felted during scouring by over agitation, if a little skilful management is brought to bear on the matter. If only this precaution is taken, manufactories that are too small to employ more than one set of scouring machines can get along very well by varying the speed of the rakes to suit the length of staple of the wool under treatment. By making this variation in the speed of the rakes, the swing-rake machine can be made to scour any kind of wool for either the woollen or the worsted manufacture.

A third method of wool scouring, introduced in 1892, is making headway towards becoming generally adopted, viz. the method of flushing the wool through a tube, trough, or cistern, by the scouring liquor, up to the squeezing rollers, without feeding a second time. The latest design of this class of machine, patented by Messrs. Petrie and Fielding of Rochdale, is so arranged that the wool is lightly held together in bulk whilst the scouring liquor is forced through it. The wool is thus conveyed evenly and regularly forward without disturbance to the squeezing rollers at a much slower speed than the scour or washing liquor flowing in the same direction. The sand and dirt fall from the wool through the brass perforated plates, and can be flushed away at the will of the operator to any independent receiver. By this method it is claimed that much less alkali is needed in proportion to the amount of wool washed. The wool is delivered clean and in a lofty condition.

During late years large quantities of wool in full grease have been imported to this country, the quick transit by steamers enabling the wool to arrive at the London sales in comparatively few weeks. It is very desirable, from a manufacturing point of view, that this practice should continue and increase, as the wool is much nicer, softer, more genial, and in every way better to manage through every stage in the manufacturing processes when it can be taken in its full grease and scoured out at one stroke; and now that wool-scouring machines are getting so common, this can be done so much more easily than was formerly possible. The scoured wool sold at the London sales has in most cases to be re-scoured before it can be dyed a bright, fast colour, as it is not clean enough for dyeing purposes; besides, most of the colonial scoured wools are scoured with soda, which is very objectionable, for the reasons already given. There is a great deal of "getting-up" about colonial scouring, the object being to make the wool as showy as possible with the least loss of weight. Nearly all such so-called scoured wools lose 10 to 15 per cent when fully scoured out for dyeing; and if not re-scoured before dyeing, the heat and action of the dye set the grease that has been left, and make it next to impossible for the carding engineer to card the wool properly, as the locks are set so fast by the action of the dye and heat upon the grease that the fibres break instead of opening and separating from each other on the cards.

It is very desirable that each manufacturer should receive his wool in the grease, so that it may be subjected to one scouring only, and that, too, under much milder treatment than it receives in the colonies.

Wool drying.—The system of drying in stoves, or upon hot iron perforated plates, is now discarded in our best modern mills, where it is necessary to have the wool soft and of good colour. Overheating in the process of drying, or more correctly speaking, allowing the wool to be subjected to great heat, and in contact with hot iron plates immediately over high-pressure steam boilers after it is sufficiently dry, is wasteful and very injurious to the fibre.

There are several machines made for drying wool, and each of these has its special claims. There is a French plan of a machine, viz. a long perforated cylinder, about 7 feet in diameter, open at each end. This cylinder is placed longitudinally upon bowls or rollers, with one end slightly elevated. It is made to rotate slowly while heated air is being blown through it. The wool enters at one end of the cylinder, and as there are a large number of spikes fixed inside pointing inwards, the wool is caught up and carried to the top and allowed to fall freely upon the spikes below, and so on repeatedly, gaining a few inches onwards at each fall, until it reaches the delivery end. This has been improved in construction, and is made by Messrs. J. and W. Macnaught of Rochdale. The principle of this machine is a good one, and wool dried in this way will always be lofty and uniform in dryness, without being scorched (see Fig. 10).

Another drying machine (Moore's patent) is made by Wm. Whiteley and Sons, Lockwood. This is a large iron house or chamber, fitted up with two trays or tables of small iron rollers, one above the other, which are caused to rotate and convey the wool through the heated chambers of the machinery for any desired length of time.

The continuous wool-drying machine of Messrs. Petric and Fielding of Rochdale is now extensively used. This is also an iron house or chamber, in which there are usually three or five conveying tables or trays, one above the other. These conveyers are strong and simple, and not likely to get out of order, as there are no wearing working parts inside the machine. The wool is blown into the machine at one end by a blast-fan, and conveyed through and through the machine, whilst the air is blown continu-

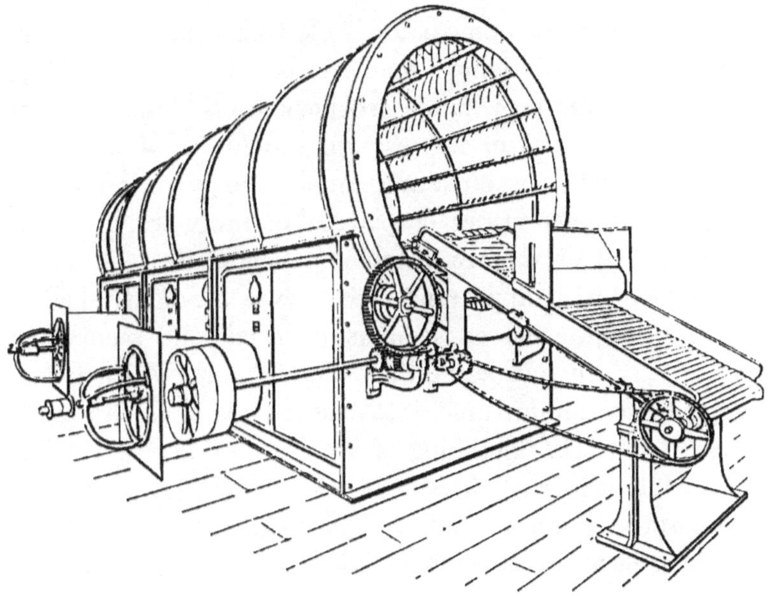

FIG. 10.—MACNAUGHT'S WOOL DRYING MACHINE.

ously and equally through the wool up to the delivery at the opposite end. The air blown through may be cold or heated to any desired temperature, and the machine is so arranged that the wool is delivered immediately when dry, thus preventing any injury by overheating. This is a very effective machine.

There are a number of table wool-drying machines in use for drying by means of hot air driven through the wool

by the use of a fan placed under the table or otherwise used
in connection with tables or scrays of various kinds.

PART VI

BLEACHING AND EXTRACTING

ANOTHER process immediately following scouring in many
manufactories of fancy woollens is that of bleaching.
For many kinds of fancy woollens it is desirable that
the white threads of the pattern should be of as fair a
white as can be obtained, and not the dingy yellow white,
which gives some styles of pattern an old or worn appearance instead of the freshness of a new cloth. To attain
this freshness of appearance recourse is frequently had to
bleaching the wool as soon as it leaves the scour, and
before it is sent into the mill. The bleaching of wool after
it has been manufactured into cloth dates back thousands
of years, but the bleaching of wool before manufacturing,
and while the wool is yet in its loose state, is only of
comparatively recent date. This mode of treatment has
not long been practised in England, but on the Continent
it has been in use for some time. Sulphur is the agent
employed, and when the material is made into a fabric or
cloth, the operation is accomplished by placing or hanging
the cloth in a sulphur chamber while in a moist condition,
but in the case of loose wool this mode is not practicably
convenient, and recourse is had to some one or other of
the sulphur solutions, in a cistern. A solution of 3 to 5
per cent of bisulphite of soda, with a like amount of
hydrochloric acid added, is frequently used. The wool for
bleaching requires to be very clean scoured, or the bleach-

ing cannot be effected. It is steeped for some hours in large cisterns, after which it requires a thorough washing in clear water and passing through the hydro-extractor.

From the illustration (Fig. 11) it will be seen that the machine is composed of a strong iron circular frame, inside of which is a circular wire cage, into which the wool or cloth is placed; the cage is then set revolving at a very rapid rate, say up to 500 revolutions per minute, and by this rapid motion the water is thrown out of the material by sheer centrifugal force against the sides of the

Fig. 11.—Hydro-Extractor.

inner part of the strong iron circular frame, which contains the revolving cage, and runs down into the bottom of the frame and out at the opening.

Formerly very heavy foundation stones, 6 to 7 tons weight, had to be placed for holding the machine fast and steady, but our illustration is of one that is self-balancing, and can be placed upon a wood floor, if necessary.

The hydro-extractor is an exceedingly useful machine, both for wool and cloth, as 70 to 80 per cent of the water can at once be got out of the wool or cloth, making the drying much more expeditious and in every way better, as a very high heat is not required, besides leaving

the wool or cloth in a more open, lofty, and better condition. It is greatly to be desired that a much more
frequent use be made of the extractor, both for wool in
the gray and for dyed wools, as nothing improves dyed
wools so much as a thorough wash, and then extracting
the water by means of the hydro-extractor. By this
means the dyed wool gets so lightened up and opened
by having the air blown through it, and by the action of
the escape of the water, that it dries with comparatively very little heat; but the great point gained is the
condition into which it puts the material for the after
operations of carding and spinning, giving it an openness and softness which it would not otherwise have,
and which is so great a help in getting the work through
these processes, and contributes in no small degree to the
best results obtainable in the two important operations.
If manufacturers would only fully comprehend the benefit
that would arise from the more general adoption of the
use of this machine in their work, its use would soon
become much more general, as it puts the wool into
a much more advantageous state for the operations
that follow, so that less waste is made, in addition to the
work being done in a more satisfactory manner. The
old method of taking the wool in its clotted state, full
of water, direct from the dye-house, and placing or spreading it upon hot plates over very high pressure steam
boilers, so hardens and cakes it that incalculable injury
is done, resulting in great and quite unnecessary waste
and very unsatisfactory and imperfect work, which could
in a very great measure be avoided by the more general
use of the hydro-extractor both for wool and woven
goods.

Extracting or Carbonising.—Before going farther it is

necessary to notice that immediately following the scouring another process has recently come into use in many of our mills, called "extracting." Extracting is a mode of chemically destroying the burr or any other seed or minute portion of straw or other vegetable matter that may be in the wool. The "burr" is a round, prickly seed that catches in the wool of the sheep as they are feeding, and imbeds itself amongst the fibres of the fleece, and is not easily disentangled. The reason for extracting is that vegetable matter must somehow or other be got rid of, as far as possible, early in the manufacturing operations. The burr is about the size of a pea, and if allowed to remain in the wool, will break up into a thousand minute particles, each one of which constitutes what is technically termed a "burl" or small mote that disfigures the cloth. Large sums of money are yearly paid in every woollen manufactory to women whose sole occupation is "burling" or picking out the small obnoxious burls, or in otherwise treating them so as to render them invisible at least for the time being. The extracting process, then, is chemically to destroy *en masse* every vestige of vegetable matter that may infest the wool, or that may lurk hidden away in the locks of wool before it enters upon the proper manufacturing processes; the earlier stages are merely preparatory. To accomplish the extracting, the wool is steeped in large cisterns containing solutions of sulphuric acid, after which it is taken and dried, in order to complete the tendering and destruction of the vegetable matter; the wool is then brought back and steeped in solutions of soda, in order to neutralise the acid that may remain in it, after which it is washed in clear water, and dried in the ordinary way. Broadly, and in a general sense, we are told that alkalies act quickly on wool, and that acids act only feebly; the acids act quickly on vegetable matter. It is chemically admitted that *caustic* alkalies and their solutions act quickly and powerfully upon wool, and to such an extent as to entirely dissolve it, if the action is allowed to continue for any length of time; but, on the other hand, we are told that acids only act quickly on vegetable matter, and feebly on animal matter, such as wool. This, then, is the popular or general chemical statement of the question, and until very recently any one objecting to it was chargeable either with ignorance or prejudice. Whatever may be the chemistry of the matter in question, practical men in our mills have long found out that if acid has been too freely used by the dyer to get up the colour, or if acid has been used for any other object, the workman can neither get through as much work, nor can he do that work nearly so well. The carder speedily finds it out, for the wool does not open out fibre from fibre as it should do, but is apt to break, very much through being tendered and matted together. The spinner also finds that the wool has been tendered, made hard and unpliable, and that he can draw neither so even nor so solid a thread, and in addition there is a great amount of breakage and waste; " it spins badly " is his short but expressive description of the situation. The weaver gets "a bad warp" and the percher a damaged piece brought before him. This is the practical outcome, whatever may be the chemistry, of saturating wool with acid. Notwithstanding all this, we are blandly told that "acids act feebly on wool." The facts of the matter amount simply to this, that *no wise, practical* manufacturer will allow any harsh treatment of his wool either by acids or alkalies. He knows perfectly well that the

road to success in the production of his goods lies in
the direction of mild treatment through every process from
the beginning to the end, and that he has the strongest
reasons for objecting at the very outset of the manufacture
to saturate his wools either with strong alkalies or strong
acids, knowing full well that the very best of their manufacturing properties are very seriously injured by these active
agents. This may be said to be looking at the matter solely
from the low level of a practical point of view; it may be
so, but that which is really true in practice, rightly understood, cannot be false in principle. The folly of the process
of "extracting" need not stop in its discussion at the
practical point of view, but may ascend higher to the
principle on which it is based, as something can be said on
the scientific side against extracting. Dr. Schlesinger of
Zurich has conducted a long series of experiments on this
subject, and published the results in a pamphlet, an
article upon which appeared in the *Textile Manufacturer* in
July 1883. The experiments were undertaken originally
in order to determine whether shoddy contained silk, flax,
and cotton as extraneous matter, and to determine approxi-
mately the proportions of each. In order fully to ascertain
the behaviour of threads of shoddy when subjected to the
action of strong alkalies and strong acids, the doctor had
recourse to the aid of a microscope possessing a magnifying
power of 350 diameters, and fitted with a micrometer eye-
piece. In these experiments there were two objects in
view: (1) The identification of the component fibres, and
(2) the quantitative analysis of the same. Having satisfied
himself as to the presence of different textile fibres, he next
turned his attention to testing the results obtained by
placing a drop of cupric acid on the object. This quickly
destroyed the silk, then began to attack the flax, leaving

the woollen fibre slightly swollen. In another sample, treated in the same manner with concentrated sulphuric acid, *the woollen dissolved*, producing a red coloration. Satisfied as to the various sorts of fibres composing the shoddy threads under examination, the next consideration was to ascertain if possible whether they consisted wholly or partially of wool that had been used before. As a preliminary step towards this object it was necessary to compare the colours, structure, and chemical characteristics of both kinds of wool, new and old. The most striking differences were found in the colours of the hairs; in the better-class shoddies the hairs were all of one colour, but in the inferior sorts they were of different hues; in some single threads of shoddy were found fibres dyed with indigo, with purpurine, and with madder, which was clear proof that they had not been dyed together, but owed their respective colours to the several fabrics of which they had once formed part. Another fact that presented itself during these experiments was the difference in diameters between the fibres of old and new wool. In the hairs of the shoddy or mungo wool the fibres contracted gradually, in other instances suddenly, expanded again, and again contracted, and in some instances continued uniform in size throughout the rest of the length. In some places the wool scales or serrations were rubbed off, in other places the hair was worn down so that those parts were below the average diameter. The behaviour of each kind of wool was noted when subjected to the action of strong potash, or soda ley, and sulphuric acid. Potash or soda ley attacked the hairs of shoddy or mungo wool more speedily than those of new wool, which had undergone no manufacture. Woollen hair treated with concentrated sulphuric acid swelled more or less, showing every symptom of animal poisoning, till gradually its primitive structure disappeared and the moment arrived when the internal band of light that had shown under the microscope disappeared also, and there remained nothing but an elongated cell covered with striæ and fissures. On the slide of the microscope was next laid crossways a hair of shoddy wool and a hair of new wool, which had been carded and divested of the grease that coats it in its normal state. A drop of concentrated sulphuric acid of 66° Beaume was placed on the point of contact of the two hairs and the respective times of touching each hair noted, a slip of glass having been laid over it, and the object observed with a magnifying power of 65 diameters, this power being sufficient for the purpose, besides having the further advantage of keeping the acid some distance from the object glass, and thus preserving it from injury. It was observed that the hair of shoddy swelled much more quickly, and to a much greater extent, than the hair of new wool, and that the shoddy hair, if dyed, almost immediately lost its colour; the discoloration gradually extended from the scales to the medullary canal until the whole assumed one uniform hue, which was coincident with the disappearance of the primitive structure. But in cases where there was no discoloration the moment of dissolution of the fibre could be fixed without difficulty, as the peculiar scaly structure of the hair was only rendered more prominent by the preliminary action of the acid. As soon as the dissolution of one of the cross hairs was complete, the time was noted, and the coefficient thus furnished indicated the relative resistance of the acid of the old and new wool that composed the object under examination. The time needed for the sulphuric acid to destroy the structure of the new wool varied greatly. Out of fifteen experiments the time needed to destroy a hair of new wool by sulphuric acid varied from 6 minutes 0·5 seconds to 1 minute and 10 seconds, and the time needed to destroy a hair of shoddy or mungo wool varied from 3 minutes and 45 seconds down to 15 seconds. In every instance the hair of shoddy offered a shorter resistance to the acid than its companion hair of new wool in the same experiment. The time required to carry each experiment through varied very much, depending on the description of wool, volume of acid, capillary current, and the varying friction and pressure exerted by the glass slips on the object under examination. In the instance where the new wool stood out 6 minutes and 0·5 seconds the hair of shoddy wool was gone in 1 minute and 45 seconds, and in another instance where the shoddy wool went soonest in all the series of tests, being only 15 seconds, the hair of new wool in the same instance stood out 4 minutes and 30 seconds. In every case in Dr. Schlesinger's experiments the destruction of the primitive structure of woollen hair by the action of concentrated sulphuric acid took place much more speedily in the shoddy than in the new wool, and this fact alone opens out a field of inquiry bearing very importantly upon the question in hand. It has already been said in the section on wool scouring that we cannot even scour the yolk and natural grease out of the wool, so as to be able to manufacture it, without going some considerable way on the road towards its destruction ; neither can we subject wool to the action of acids, however slightly, without travelling the same gloomy highway. We have long practically had to contend with the pernicious effects of the too free use of acids upon wools, and the scientific reason " why " of those effects is being brought out more clearly year by year. We must not omit to draw the conclusion from another effect that appears in these experiments ; the old wool fibre offers feebler resistance to the acid than the new, because the old wool fibre has had its constitutional strength undermined and reduced by having gone through the manufacturing processes previously, one or more times, and run the gauntlet of acids and alkalies, having had to suffer the effects of their weakening action. This being so, need we wonder or feel any surprise that its constitutional strength is reduced, and that it has become feebler and can interpose far less resistive force when brought face to face again with its old deadly enemies ? One thing is clear, both from practical and scientific research, that if we would keep our wool material sound, soft, and pliable, and if we would produce and sustain the beauty of colours, we must jealously guard it against the action of alkalies and acids at every possible point throughout the whole course of its manufacture. Few persons have any conception of the marvellous delicacy of the structure of the wool fibre and how extremely susceptible it is to injury through harsh treatment, how soon its fine scaly surface can be fatally affected, its milling or felting properties destroyed, and its general strength reduced by the destructive action of acids and alkalies. If this new-born process of "extracting" must be persisted in, contrary to scientific and practical knowledge, let it be applied to only a part of the wool, say to about 10 per cent of the whole ; that is, to the very burriest portion. In sorting the wool let the very burriest parts of the fleece be cast aside for extracting, and the bulk left to be dealt with by the burring machine. There is one rule that invariably ought to be followed by all manufacturers, *never to do anything chemically that can be done mechanically.* We shall have to speak of this aspect of the subject when we come to discuss the burring machine.

PART VII

WOOL DYEING

THE art of dyeing is of great antiquity, and the vicissitudes of its progress are deeply interesting. No dyes are probably so old as the *Indicum nigrum* or Indian ink and the *Indicum purpureum* or indigo. It is said that the Phœnicians and Egyptians learned their use from the native seat of their production—the East, and the great aptitude of the ancient Greeks for every art soon induced them to borrow a knowledge of their uses from the Phœnicians, and that the Romans in turn were instructed by the Greeks. Authorities tell us that in the wreck of civilisation which followed the break-up of the Roman power many of the best and most useful inventions perished, and the knowledge of physical science generally fled back again to the East, and that the Arabians for a lengthened period became its chief depositaries. In the twelfth and thirteenth centuries the Jews were the dyers of Europe; they "had learned this art in the East, where it was carried on in many things to a degree of perfection which the moderns have not been able to attain," and it was from this source that Florence and Venice, in the heyday of their manufacturing prosperity, learned to give to their cloths such beautiful and durable colours. In this wise a knowledge of the art of dyeing has come down to our own times, and has been brought to such a degree of perfection as our ancestors had, so far as historical evidence goes, no conception of. But it may be said here that in our modern times we have attained to a science of dyeing, in addition to the art, though much yet remains to be done in this direction.

The science of anything is the fixed principles that underlie its foundations as an art; from these fixed principles are deduced certain practical rules, and the application of these rules constitutes the ART of the industry in question. Dr. Whewell in his *Inductive Sciences* says, "Art in its earlier stages is anterior to science; it may afterwards borrow aid from it." In ordinary practical life this is so, as mankind sees effects before discovering causes; it is the presence of the effects that ordinarily induces men to search out their causes. Still, in the true order of succession causes go before effects, and underlie the foundations of all; so in like manner science goes before and underlies the foundations of every art, though it is unseen. Science gives principles from which art obtains its rules. If our knowledge of anything be merely an accumulated experience, then the art based on such a foundation is said to be empirical, having no scientific foundation.

Dr. Thomson says in his *Laws of Thought*, "The distinction between science and art is that science is a body of principles and deductions to explain the nature of some object matter. An art is a body of precepts with practical skill for the completion of some work. *A science teaches us to know, an art to do.*" We ought therefore to seek to know, and we shall work more intelligently, and consequently more efficiently. Science and art, and what pertains to each, are in no department of modern industry more intimately linked together and more strikingly illustrated than in dyeing. Here, if anywhere, the science of *to know* and the art of *to do* must go hand in hand, if complete success is to attend the issue. The practical dyer has to ply his vocation under so many varying conditions and such differences of water in different localities, and the constantly varying qualities of dye-wares by which he has to obtain his colours, that unless he has some scientific knowledge of his profession the quality of his work will very materially vary from day to day, notwithstanding his having bestowed upon it his best efforts and his unflagging attention.

And further, again, the practical dyer may in his every-day work as an artist carry out very imperfectly the scientific principles of his profession, yet the imperfect skill of the artist, or the imperfect application of that skill, ought not to discredit the fixed principles that never vary, and that ever underlie the foundations of his art. The artist seldom reaches the perfection of his own ideal, and the artist dyer frequently fails to reach the scientific ideal of perfection in relation to some certain specific colour; therefore in practice we must always allow a margin for the imperfections of the artist in his attempts to reach the mark of scientific perfection that is set before him. The practical application of anything is but its art, and the science of it is most frequently below the surface; there *is* a science of it, if we will only dig down deep enough to find it, and the science of dyeing lies deep down in the chemistry of its operations. These remarks apply not only to dyeing but to many other of the intricate and complicated operations of the woollen manufacture; there is a most urgent need to appeal to first principles in order to check the tendency to false courses in our attempts at improvement. And notwithstanding a great amount of prejudice in the minds of a part at least of the manufacturing community, as there are many who yet contend that it is impossible to obtain the desirable effect, that is, both the chemical theory and the practice combined in one person, still the chemistry of dyeing comes more and more towards the surface of daily manufacturing life as the years go by, and will continue to come more rapidly forward as technical knowledge increases, till, finally, every operation of the dyer's art will be traced backward and downward to its scientific foundation.

Wool, as well as the other animal fibres, is dyed in its loose state, after having undergone the operation of scouring, which we have just noticed. It is also dyed in its manufactured state when made into cloth, and in its semi-manufactured state of spun yarn, and in each of these states or conditions the mode of operation has to be modified in accordance with those conditions. Formerly the manufacture divided itself into two great parts—"wool-dyed" and "piece-dyed" goods; but since the rise of the fancy woollen manufacture much of the material is dyed in the wool or loose state, before it is manufactured into either yarn or cloth. Much also of the material is now dyed in its semi-manufactured state of yarn or in the thread, as the dyeing cannot be effected when manufactured into cloth, where a woven fancy pattern composed of a great variety of colours goes to make up the fabric. In all such cases the material must either be dyed in the loose wool at the outset, or it must be dyed midway when it has reached the point of spun yarn or thread, which is immediately before it is woven with a multitude of other colours to form a pattern. Where there is a woven pattern upon the fabric or cloth, but where the pattern figure and the ground of the cloth are intended to be the same colour (drab, for instance), then the fabric may be "piece-dyed," as in the olden days. In "piece-dyes" of this kind threads of dyed yarns are frequently introduced in the warping, and so used in stripe figuring as to bring out some very beautiful effects. As an instance of this take the drab shade just named and

introduce some drab yarns into some kind of feathery
stripe, and then, when the cloth has become "piece-dyed,"
the threads in the stripping that were already drab take
on deeper shades, and some rich effects are produced by
a gradation of shades. Of course, this idea can be carried
out in many other colours of piece-dyes as well as drab.
While at this point it may be stated that threads of vegetable fibre are sometimes introduced in this way either for
spotting, stripping, or checking, in which case the vegetable
material will not take dye with the animal, and when the
cloth has been "piece-dyed," the pattern appears. From
the instance just cited it will appear that wool does
not stand in the same relationship to colouring matters as
vegetable fibres do, but takes on colour much more readily
than the vegetable products, and hence there is a great
difference in almost all the operations of dyeing through
which it has to pass. Wool has a much stronger affinity or
attraction for many of the colouring matters than vegetable
substances have, and consequently there are more substan-
tive colours in wool dyeing than in any vegetable material
or texture, so that in many cases there is no need to resort
to those dyer's friends, the MORDANTS. Perhaps there are no
expressions that "bother" the young student more than
these "affinities" and these "mordants," and no expressions
are more difficult to explain in such a manner that a
student may thoroughly understand them. We will therefore
try to illustrate their meaning. A comes into contact with
B and likes him, and B in turn has an affection for A; the
two have a natural liking for each other's society and enjoy
it. Natural liking of things for each other in the chemical
world is "affinity." Then further, though A has a strong
"affinity" or natural liking for B, he has no such spon-
taneous natural liking for C, but a like affection exists

between B and C; yet though A cares little for C in particular, but is fond of B and B is fond of C, therefore the little society hangs together in harmony through the influence that B has upon each of his companions that sit right and left of him.   B occupying a position analogous to this in the chemical world, the dyer would at once see that he stood in the exact position of a mordant, and in chemistry could act as one.   A mordant, then, in dyeing is a third factor or agent that is called in to unite two other factors, such as wool and some one of the colouring matters that would not unite of themselves, because they have no "affinity" or natural liking for each other, and can only be united through a "mutual friend" or mordant that has an "affinity" on the one hand for the wool, and on the other hand for the colouring matter required to be used to produce the desired colour.   Mordants, therefore, are things used in dyeing that have a power or capacity in them of "double affinity," one on the right hand, and another on the left.

Then further, again, the dyer varies his art by producing a variety of colours from the same base, employing different "mordants."   Take as an illustration our well-known colouring agent "madder."   Madder yields a *rose hue*, if acted upon by one mordant; but if a certain other mordant is brought to bear upon it, the colour produced is a *dark brown*.   Take a third mordant and the resulting colour is *dark red*, and a fourth mordant may be used and the colour brought forth is *black;* this proceeding may be reversed, and the mordant employed be the same, but employed in connection with a great variety of colouring agents.   The mordants, it will be seen, play a very important part "for weal or for woe" in the operation of dyeing, and on their rightful *use* or their *abuse* depends much of the misery and worry of manufacturing life; and not

H

only this, but when the fabric has at last found its way to the wearer, on the rightful use or abuse of the mordant very much depends whether the colour of the garment is durable or otherwise. Bearing upon this point a very long chapter might be written *on the use and abuse of bi-chrome* alone as a mordant, so greatly is it affecting our manufacturing welfare. For there is, perhaps, not in the whole range of wool dyeing any agent more subject to *use* and *abuse* than this, and now that it has become so cheap in comparison to what it was formerly, and to all appearance is likely to become cheaper than ever before known, perhaps a few words may be said with benefit. Bichrome is in many small dye-houses almost the only mordant, and resorted to in almost every case where a mordant is necessary; yet there need be no hesitation in saying that it is doing, and has already done, an immense amount of harm through its indiscriminate use. No person who mixes in society in the market towns, and that notices the various tones of black, can have failed to observe that its effects are very palpable in the garments on the back of many a wearer of worsted coatings, in the shape of that well-known greenish-looking hue that makes a cloth look old before it is half worn. Let us for a moment refer to another bearing of this important matter, that which especially affects that noted branch of the woollen trade, the black cloth. A piece is sent to the dyer with strict injunctions that it must be "woaded," that it must have a ground of indigo put upon it for making the colour of the cloth or wool more durable and lasting in the wear. Assuming now that the grounding of the colour has been with indigo, the next process is the "preparing" of the wool or cloth after leaving the indigo vat to cause it to have affinity for logwood, camwood, etc., to complete the black dye; and here in ninety-nine cases out of every hundred bichrome is the agent used for the mordant, and the result often is that fully one half of the indigo put upon the piece or wool is uselessly destroyed instead of being left on the wool or cloth to effect its legitimate purpose of supporting and prolonging the stability of the colour; it is destroyed in this mode of "preparing," and the cloth or wool in many instances had better never have been put into the indigo vat at all, the operation resulting in a waste of time, waste of indigo, and injury to the material. For proof of this statement let any manufacturer ask for or take a pattern from a piece or wool that has just come out of the "woad" vat, and then let him procure another pattern from the same piece or wool after it has undergone the operation of "croming" or preparing in bi-chrome in the usual way, and the difference will speak for itself, and is worth noting. A return to the old mordants of iron, tartar, and alum in cases where indigo is used would in many cases be an advantage to all concerned, but more especially to the wearer. If merchants only knew a little more of the ill effects of the use of this mordant they would insist upon bichrome being left out and other mordants substituted, especially in better-class goods where stability of colour is of the first importance. Of course, in the use of bichrome we are free to admit that there is an immense amount of variation, scarcely any two dyers using the same amount for producing the same effect. But we would strongly urge that for the welfare of our wool manu-facture the indiscriminate use of an agent so pernicious as a mordant should be reduced to a minimum. It is impossible to state in exact terms the amount that ought to be used in all cases, as there are different circumstances attaching to certain cases which render it necessary in many instances to treat each case on its own merits. But it may be broadly stated that there are few cases that need exceed 2 to 2½ per cent of the weight of the material to be dyed where black is the colour desired, providing that the goods are washed clean from all alkali, and a little acid used, so as to enable the full effect of that quantity to be obtained and properly fixed. So much as 7 or 8 per cent of the weight of material to be dyed has been used of bichrome, to the serious injury of the material and of the colour, and consequent disappointment to the wearer. But this is not a work in which to discuss at further length the bearings of this matter, suffice it to say that a return to the old mordants would in some cases be a great advantage to the woollen trade generally, as the old mordants have less tendency to destroy, and work infinitely better with indigo, and as a consequence the woaded colours produced on the old basis are much more durable and satisfactory. To the scientific colourist it may be unnecessary to say that there are, practically speaking, only three primary colours, and that the three ancient primaries indigenous to Western Europe were "madder," "woad," and "weld," but to the elementary student some such reminder is necessary. The madder yields the red dye, the woad the blue, and the weld or wold the yellow, all of which are fast durable colours, to scouring and milling, but not to light. Madder (*Rubia tinctorum*) is largely cultivated in the south-west of Europe, the district round Avignon, in France, producing large quantities. Woad, the *Isatis tinctoria* of botanists, is indigenous to most parts of Western Europe. It is still cultivated to a considerable extent in France, but its cultivation in England is now narrowed to Lincolnshire. In Germany it was also formerly very largely produced in the district around Erfurt, which, along with the surrounding places, were denominated the "woad towns." The leaves are twisted off from the plant two or three times during a season, and the manufacture of the woad is carried on during the winter by the leaves having lime and urine sprinkled amongst them, and then being piled up into a heap to ferment, being turned over with a shovel at intervals to prevent excess of fermentation, to which the mass would otherwise be liable. When fermentation ceases, the woad is ready to be packed for the market. Weld or wold (or, as it is sometimes called, the dyer's weed) is a native of Britain, Italy, and other parts of Europe, and is cultivated for its stalk, flowers, and leaves, which are employed in dyeing yellow, from which arises its botanical name *Reseda Luteola*. The wold is still preferable where that soft, delicate canary or lemon yellow is required to be produced to stand milling, and the colour stands the action of alkalies well without turning brown. From the above three elementary, cardinal, or primary colours the dyer eliminates the secondary, tertiary, and other combinations, and these produce an almost infinite variety of shades.

The Woad or Indigo Dye.—The ancient woad plant (*Isatis tinctoria*), from which our forefathers derived their permanent blue, is really a species of the indigo family of plants, and the colouring matter derived from it is essentially indigo, just as is the colouring matter derived from the richest variety of the "Indigofera" of Bengal. Indigo blue is not found in any plant in its natural state, but is a manufactured article, and is obtained by subjecting the leaves of the plants, when reaped, to a peculiar kind of fermentation. The percentage of colouring matter yielded by the woad plant of Western Europe is small indeed compared with that of the "Indigofera" of India. The product of colouring matter yielded by the family of indigo plants varies with the climate; it is low in the temperate climate of Western Europe, where vegetable life is weak, and high in hot climates where vegetable life is energetic and vigorous. The colouring matter varies in this class of plants from a very small percentage in Western Europe to from 60 to 70 per cent in the richest variety of the "Indigofera" of Bengal. Indigo found its way to Europe overland through Alexandria long before the route to India was discovered by the Cape of Good Hope, and when first brought here furious opposition was raised against its use by the woad growers, and a little was secretly added to the woad vat to heighten and improve the colour. It was ultimately found that a combination of the two dyes was more penetrating, and gave a superior and more permanent colour than when used alone, so that little by little the quantity of indigo was increased till it reached the condition, or mode of working the woad vat, in which we now have it. The Yorkshire 7-feet woad vat is composed much after the following manner:—

<pre>
560   lbs. Lincolnshire Woad.
 18   ,,   Bran.
 22   ,,   Slaked Lime in powder.
 2½   ,,   Madder.
 24   ,,   Indigo, finely ground till perfectly smooth.
</pre>

Other formulas are given that differ in some points from the above, different woadmen using slightly different quantities of wares in their make-up of the vat, but the above will suffice for our purpose. The woad cut into small pieces is cast into the vat, which is then filled with water, after which it is heated by steam to a temperature of 130° to 140° F. The vat has now to be stirred every fifteen or twenty minutes for five or six times, and then left till the following morning. On the morrow, when the woad has become soft, the vat has to be well raked four or five times at intervals, and the bran, madder, indigo, and half the lime are cast in, and the vat covered over and left till the following morning. On the second morning after the setting, if fermentation has commenced, a slight froth will rise to the surface, after the bottom has been gently stirred with a rake, and the liquid when disturbed will begin to be of a greenish yellow, interspersed with blue veins when stirred with the hand, and the coppery blue scum called "flurry" begins to appear on the surface and the smell of the vat begins to be somewhat sour. If these symptoms appear, the vat is said to be going on all right, and a quart more of the lime-powder is added and the compound well raked up. It must be borne in mind that the woad vat, when in a healthy condition, is one of continuous fermentation ; of this the young student should make a note, and if at this stage no signs of fermentation appear, the additional lime is withheld and more time allowed for fermentation. If the vat comes on satisfactorily to maturity a quart of lime-powder is added about every three hours, or thereabouts, stirring up well at each addition. The coppery blue "flurry" should have increased in quantity, and the liquid when stirred should show blue streaks, or veins, amongst the general appearance of greenish yellow. On the third morning after the day of setting, if the vat has come forward all right, it should be in a condition to commence dyeing. But the whole process of managing the woad vat is a matter involving the greatest niceties of judgment on the part of the woadman, the whole of his five senses being called into action ; except he has a keen sense of smell he cannot make a good woadman ; the experienced woadman can detect by his sense of smell imme-

diately he enters his dye-house if a vat is going wrong;
he also uses his sense of sight sharply to detect every varying symptom in the appearance of his vat, and bends over
and listens to the fermentation with all the delicate and
refined sense of hearing possible ; he also uses his sense of
touch by noticing if the vat feels slippery or harsh to the
fingers, and of taste to aid him in his judgment.   We are be-
traying no state secrets in endeavouring to give the student
some account of the woad vat, nor are we in the least
depreciating the value of the services of the woadman dyer,
but far otherwise, as nothing short of the most mature
experience can manage the vats ; there is no room here
for any kind of " raw-headed " amateurs.   The woad vat is
in perpetual fermentation, day and night, week in and week
out, and its efficiency as a dyeing agent depends on the
fermentation being kept, as it were, on a balance at one
particular point.   If the fermentation flags and falls short
of this particular point, then more vegetable matter has
to be added to quicken and increase the fermentation ; while
again, on the other hand, should fermentation become too
active, lime must at once be added to check the excess of
it ; should lime not be added, the fermentation without
check would pass on to the putrid stage or condition, and
the vat would " run away," as it is technically called, and
the indigo would be lost or wasted.   From this it will be
perceived that the most constant care and attention,
coupled with the most consummate skill on the part of the
woadman, are called into requisition in order to keep the
fermentation on the delicate balance absolutely needed in
the proper management of a woad vat, and no description
can accurately convey to the mind the amount of discriminating judgment that is requisite to meet the ever-varying
peculiarities of a woad vat.   Youth and age must stand

elbow to elbow to acquire a knowledge of its ever-varying symptoms, and to learn to quickly detect any approaching change. All the science in the world, apart from experience, cannot manage a woad vat; but experience and science combined, and mutually helping each other, place it under complete control. For keeping up the strength of the vat, about 15 lbs. of indigo, more or less, according as it is worked in dyeing goods, is needed per day, and more lime or bran added daily, as each may be required, to keep the fermentation on its balance. Whether the vat is doing any dyeing work or not, it requires raking or stirring and repeated attention every day of its existence, as it is constantly undergoing change, and that change has to be as constantly watched by the experienced *eye* and *hand* and *smell*, in fact, by the whole faculties of the woadman, to ascertain whether the change tends towards excess of fermentation on the one hand, or to the want of it on the other; the smell changes very rapidly if too much or too little lime is present, and so in like manner in regard to excess or want of vegetable matter.

When the vat is ready for work, an iron hoop, to which is attached a trammel net, with 4 or 5 inch openings, is suspended some 3 feet down in the liquor when for piece-work; when for wool, a net with half-inch meshes or openings is used to prevent the goods mixing amongst the wares in the bottom of the vat. When the vat is employed on piecework, the cloth, if worked by hand labour, is moved about in the liquor by an implement technically called a "hawk"; if by machinery, then a series of rollers are used called a hawking machine, care being taken always to keep the goods beneath the surface, or the colour would be uneven. Woad vats are of various strengths in a dye-house, just as they are new or of advanced age. When the piece of cloth is for a *full blue*, it is worked from one to two hours in a vat, and then removed to a new vat to fill up the colour. If the piece or wool is for woaded black, the goods are placed in a weak vat for twenty minutes or more, according to the depth of blue or woad required.

From the above the student will now be in a position to understand something of the woad vat, and he will also understand something of its constitution and working. It is well worth the student's best endeavour to comprehend something of its intricacies. Indigo dyeing is one of the most interesting and curious of any of the operations in the whole range of the wool industry.

The Black Dye.—A good black has always been a desideratum. As far back as the middle of the fifteenth century, when Venice stood high in the art of dyeing, a good black was so much prized that the dyers in black and the dyers in colours formed separate guilds—the black being deemed of so much importance as to merit a separate guild for the cultivation of the knowledge of it ; and as it was in the middle ages so it is now, we are still crying out for good blacks, and the cry is met by the contemptible habit of putting a 4s. per yard dye upon 12s. per yard cloth ; the old ambition to produce a good durable black seems to have died out. The following is a present-day black, with part of the chrome held back :—

BLACK—108 lbs. clean Wool.

For preparation.
$\left\{ \begin{array}{l} 3\frac{1}{2} \text{ lbs. Bichrome.} \\ 2 \quad \text{,, \quad Ground Argol.} \\ 2 \quad \text{teacupfuls Sulphuric Acid wash-out} \\ \quad \text{of this preparation.} \end{array} \right.$

For filling up.
$\left\{ \begin{array}{l} 50 \text{ lbs. Common Logwood.} \\ 16 \text{ lbs. Camwood,} \\ \quad \text{and finish with 8 oz. Copperas.} \end{array} \right.$

There is a fraction more chrome in the above recipe than the quantity advised in the preceding remarks, but this is a practical recipe, and is in daily use, and the quantity of bichrome could no doubt be lessened, provided due care was taken to have the wool thoroughly clean scoured—a matter in which we are far from having reached perfection. The practical dyer, therefore, has to adjust his quantities of dye-wares a little in excess of that which is absolutely necessary, in order to have a margin by which to protect himself against the carelessness and short-comings of those around him. It will be observed that this is not entirely a logwood filling up ; there is a portion of camwood, and why ? The camwood is there for a twofold purpose ; first, because it gives a full-bodied, durable colour, and, in the second place, because it furnishes a red set-off against the very objectionable tendency to greenness on the part of the chrome. Further, it will be perceived that the dye is topped with copperas, which is another safe-guard against the fell tendency of the chrome. This tendency of the chrome towards green has in all cases to be provided against when stability and durability of colour have to be kept in view. Altogether this is a fairly representative and substantial common black, a more serviceable colour than many of the so-called "woaded blacks," which in too many instances are nothing more than "a delusion and a snare." The only fault in this black is that there is a tendency to bleed red in the washing off, thereby staining in a slight degree the light shades that are used in connection with it in fancy cloths ; but this inconvenience can be got over by the use of a little common salt in the final washing off, and by the further precaution of never allowing the washed goods to lie long in that state, but to put them through the hydro-extractor at once, and then get them through the tentering machine with the least possible delay. A great deal of the mischief is frequently done in forcing the dyer to get up his colours so as not to "bleed," but to "wash off clear." If, for the sake of peace, the dyer does make them wash off clear, it can be done, but only at fearful cost to the colour. But as nothing else will satisfy unreasonable, uninformed people, the dyer often submits, and makes the goods wash off clear by chroming them on the top, finishing with chrome. So strong is the tendency of the chrome to go green as soon as it is exposed to the atmosphere that there have been instances known where garments made from goods topped with chrome in this way have perceptibly shown a tendency towards greenness before they left the hands of the tailor. Such an amount of damage done to the durability of the colour in order to avoid a moment's inconvenience in the washing off after the final cleaning of the goods for the "finisher"! Can folly go farther than this? However, the goods *wash off clear*, and serene ignorance is satisfied. But what about the colour? Well, about the colour, that is simply chromed to death; more people are to blame for that than the dyer; he has been urged, and even driven, to it. Chrome in the bottom of the dye and chrome on the top; is it a matter to be wondered at that the dye does not stand the action of the atmosphere? The only rational ground for wonder would be if such a colour, sandwiched in chrome, could stand the atmosphere for a single day without going. If the colour is to be at all durable, the mischief-working chrome must be buried as deep in the bowels of the wool fibre as it is possible to get it, and the body-weight of all the dye piled upon it to keep it out of sight, and finished with a good body of colour, inclined to red, such as is given by the camwood and the iron of the copperas.   Only by some such course as this can the chrome be kept under and the colour made durable.

There is another way of topping up a black on loose wool, and that is by using tin solutions.   The camwood may be wholly, or in part, withdrawn and logwood substituted in its place, and the purple bloom obtained by the tin topping. But as tin acts somewhat fiercely on wool, it is necessary to apply it with caution, and instead of throwing on the tin with the pan at full boil, run off say a quarter of the liquor, and fill up the pan again with cold water till the temperature is let down to about 180° F., then throw on the tin slowly, meanwhile handing up the wool and keeping it in active motion for a short time.   If the tin be thrown on with the pan at full boil, it will knit up the fibre of the wool and make it difficult to open when it reaches the carding room. Whatever mode of topping a black be resorted to, provision must always be made for keeping down the greeny tendency of the chrome by a balance of red or purple tint; otherwise the colour cannot be satisfactory or durable to the wearer.

There has been a craze of late for thin blue-blacks, and these kinds of thin watery shades can only be obtained by using the very commonest of logwood; a first-class Honduras logwood gives too much body of colour and too much rich bloom.   It requires no great amount of wisdom to find out the strength and position of a colour that is produced by the lowest kind of logwood and that is chromed beyond all reasonable bounds.   There used to be less complaining of "poor blacks" when good Honduras logwood and a little camwood were used in the filling up, and when people had the good sense not to ask for "blue-blacks" other than those on a good woaded foundation; but the more durable colour-producing agents are now too frequently left out under the plea of "cost," and cheaper woods, such as sanders and barwood, are made to do the best apology duty that they can in place of those that are reliable and more durable; but there is no cheapness so dear as inferior colours in the woollen manufacture. The colour put upon a fabric or cloth should bear such a relative proportion of value to the other sections of the manufacture as to stand its ground sufficiently long till the fabric can be said to be fairly well worn out. This has not been the case during recent years. The colour, in better-class goods especially, has not been in proportion to the gross value of the cloth, and when the colour is gone the general wear of the garment is at an end.

Aniline is now frequently used for the topping of black and blue dyes, but this, like the chrome topping, should be discouraged in every possible way as being not durable. The discovery of the aniline dyes has given a greater resplendency to wool colours than was ever before known, but the anilines are not available to any material extent for the woollen manufacture proper, that is, for goods that have to undergo the process of scouring and milling after they are woven. The wasting "wear and tear" of alkalies, acids, dirt, and heat inseparable from the manufacture either completely destroys the colour or takes off the brilliancy to such an extent as often to render the colour useless. Indeed, it may be laid down as an axiom that the more brilliant the colour, the less durable it will be found to be. This test-truth holds good all the way down the colour scale, even down to indigo, which is more brilliant than its twin-brother, the colouring matter of the woad, but less durable, and hence indigo can acquire durability and be used to most advantage in combination with woad, both as to durability and economy of use.

Before entirely leaving the subject of the black dye it is necessary to say something about tests for indigo on woaded blacks, so that those dyers who are turning out good, honest work may have the full credit of it, and in the present state of the dyeing art it may be said that nitric and muriatic acid tests are not at all reliable for use by the general public. The nitric acid, for instance, is taken and dropped upon a cloth said to be "woaded," and indicates orange; but orange now will often be the indication if dropped upon a cloth wholly dyed with logwood, as there is so much touching up and trimming to dodge the old tests by making the orange tint to approach as near as possible to the desired hue, therefore the old test is misleading. This azure film of indigo at the bottom of the dye of the "woaded" cloth certainly gives a slightly different tint to the orange indication, but the difference is only discernible to the well-trained eye of a good dyer. The merchant and the wearing purchaser cannot be expected to perceive such niceties of difference where in many cases the well-skilled dyer finds it necessary to strip the colour to the foundation before any trace of indigo can be found. The "woading" of blacks in too many instances has become so much a formality, the indigo put on is so slight, that the old tests are no longer of any service, and reliance can no longer be placed in them. The only really reliable method now is to strip down the colour to its naked foundation, whether wool or cloth, before venturing to say what amount of woad or indigo there is upon the wool or cloth in question, and this cannot be done quite so quickly and conveniently as dropping a few drops of acid upon a portion of wool or cloth; but in a matter of so much importance it is in many cases worth while to spend a short time in probing the matter to the bottom, and getting to know how much indigo you get for your money. The following instructions will be sufficient in many instances for any ordinary person to satisfy himself as to the presence or otherwise of "woad" upon the wool or cloth for which a woaded price has been paid. Procure a porcelain dish or glass beaker and a piece or two of small glass rod with which to stir and take out the bits of pattern, then mix together methylated spirit and hydrochloric acid in equal parts, add eight or ten parts of water to one part of the mixture, heat gently in the dish or beaker, and put in the bits of wool or cloth to be tested for the presence of indigo blue or woad. After ten minutes take out the bits of patterns and put in fresh liquor, into this put the bits of patterns, and heat up as before; and if need be repeat this process a third time, when the woods of the dye will generally be all, or nearly all, stripped off and only the indigo blue or woad left; but should the blue not be thoroughly bared, repeat the operation a fourth time, which will be sufficient, providing no artificial blue has been employed. Here we must leave the black dye and proceed to give illustrative instances of other colours.

DARK BROWN—300 lbs. Wool.

For the preparation. { 3 lbs. Bichrome.
                      1 pint Sulphuric Acid.
                         Boil 1 hour.

For filling up. { 40 lbs. Madder.
                  36 lbs. Fustic.
                  60 lbs. Camwood.
                   3 lbs. Logwood.
                  Boil in this for 2 hours, then add
                  2 lbs. Copperas, after which allow
                  to stand for a time before letting
                  off.

This is one of those colours that can be greatly

improved by " woading " when the material is intended for better-class goods, were it not that there is that mono-syllable " cost " everlastingly interposing its veto; still, notwithstanding this, the colour ought to be so prepared or grounded when intended to be used for first-class goods. The more composite a strong colour like this can be made, the more pleasing it is to the eye and the more durable it is to the wearer.

LIGHT OLIVE—125 lbs. Wool.

For the preparation.
$\begin{cases} 1\frac{1}{2} \text{ lbs. Bichrome.} \\ 1 \text{ lb. Blue Stone.} \end{cases}$
Boil 1 hour.

For filling up.
$\begin{cases} 80 \text{ lbs. Fustic.} \\ 2\frac{1}{2} \text{ lbs. Madder.} \\ 3\frac{1}{4} \text{ lbs. Camwood.} \\ \frac{3}{4} \text{ lb. Logwood.} \end{cases}$
Boil in the above $1\frac{1}{2}$ hours.

LIGHT DRAB—250 lbs. Wool.

For the preparation.
$\begin{cases} 1 \text{ lb. Bichrome.} \\ 1 \text{ pint Vitriol (Sulphuric Acid).} \end{cases}$
Boil 1 hour.

For the filling up.
$\begin{cases} 1\frac{3}{4} \text{ lbs. Madder.} \\ 1 \text{ lb. Fustic.} \\ 4 \text{ oz. Shumac.} \\ 1\frac{1}{4} \text{ lbs. Camwood.} \\ 10 \text{ oz. Logwood.} \end{cases}$
Boil 2 hours, then let off.

LAVENDER—100 lbs. Wool.

2 lbs. Tartar.
5 lbs. Alum.
16 lbs. Logwood.
$\frac{1}{2}$ oz. Indigo Extract (dry).
Boil 1 hour, then give $\frac{1}{2}$ lb. Copperas, after which allow $\frac{1}{2}$ hour longer, and then let off.

All these illustrations of common pan colours will admit of being woaded when intended for high-class goods, and would be greatly improved by it, except the light drab, which is too light a shade to admit of carrying a woaded ground.

The above are only short examples of obtaining the afore-named colours. Within the last few years considerable changes have taken place in the methods employed, and the colouring matters from the so-called alizarine group have been gradually coming to the front and taking the place of the old dye stuffs, and in many cases with advantage, as there is greater freedom from the small particles of woody fibre that were always left more or less entangled in the loose wool, and consequently were liable to be injurious in the further processes of carding and spinning; besides this there has been a decided reaction against the more fugitive colouring matters that came into use with the advent of the aniline colours, and a range of colouring matters are now put into the dyer's hands which, when properly employed, will undoubtedly be to the advantage of the wearer and the public generally. As an example of this class of dyeing we give the following, viz. :—

LAVENDER—240 lbs. clean Wool.

OLD METHOD.		NEW METHOD.	
		Mordant.	
Alum	. 5 lbs.	Bichromate of Potash . $3\frac{1}{2}$ lbs.	
Logwood .	. 4 ,,	Tartar . . $1\frac{1}{2}$,,	
Cudbear	$\frac{3}{4}$,,	*Dye.*	
Tartar	. $2\frac{1}{2}$,,	Gallocyanine . $2\frac{1}{4}$ lbs.	
		Acetic Acid . 2 pints.	
		(For a bluer tone use a little Alizarine Indigo Blue S.E.P.)	

MEDIUM DRAB—240 lbs. clean Wool.

OLD METHOD.	NEW METHOD.
Mordant.	*Mordant.*
Bichromate of Potash . 5 lbs.	Bichromate of Potash.
Dye.	*Dye.*
Rasped Fustic . 9 lbs.	Chip Fustic . . 28 lbs.
Madder . 12 ,,	,, Logwood . . 1 ,,
Cudbear . . 1 ,,	Alizarine Red W.R. 2 ,,
Logwood . . $\frac{1}{2}$,,	

DARK BROWN—240 lbs. clean Wool.

Mordant.	*Mordant.*
Bichromate of Potash . 7 lbs.	Bichromate of Potash . 7 lbs.
Dye.	*Dye.*
Chip Fustic . 90 lbs.	Chip Fustic 120 lbs.
Madder . 10 ,,	Logwood . . 12 ,,
Camwood . 10 ,,	Alizarine Red W.R. 10$\frac{1}{2}$,,
Sanderswood . 45 ,,	Boil 1$\frac{1}{2}$ hours, then sadden
Logwood . 10 ,,	with Copperas, 6 lbs.
Boil 1$\frac{1}{2}$ hours, then sadden	
with Copperas, $\frac{1}{2}$ lb.	

OLIVE—240 lbs. clean Wool.

Mordant.	*Mordant.*
Bichromate of Potash . 6 lbs.	Bichromate of Potash.
Dye.	*Dye.*
Chip Fustic . 64 lbs.	Chip Fustic . 64 lbs.
Logwood . 10 ,,	,, Logwood . 10 ,,
Madder . 12 .,	Alizarine Orange W. 3$\frac{3}{4}$,,
Camwood . 6 ,,	Boil to pattern.
Boil to pattern.	

DARK GREEN—240 lbs. clean Wool.

Old Method.		New Method.	
Mordant.		*Mordant.*	
Bichromate of Potash . 5 lbs.		Bichromate of Potash . 7 lbs.	
Oxalic Acid . . $2\frac{1}{2}$,,		Tartar . . . $2\frac{1}{2}$,,	
Dye.		Oxalic Acid . . $2\frac{1}{2}$,,	
Chip Fustic . 25 lbs.		*Dye.*	
,, Logwood . 30 ,,		Coerulein W. . 30 lbs.	
		Aliz. Blue W.H. 18 ,,	
		Acetic Acid . 1 gallon.	

DARK BLUE—240 lbs. clean Wool.

Mordant.		Mordant.	
Mordant.		*Mordant.*	
Bichromate of Potash . 5 lbs.		Bichromate of Potash . 6 lbs.	
Alum . 20 ,,		Tartar . . 3 ,,	
Oxalic Acid . $2\frac{1}{2}$,,		*Dye.*	
Dye.		Chip Logwood . 45 lbs.	
Chip Logwood . 120 lbs.		Gallocyanine . 10 ,,	
Cudbear . . 5 ,,		Alizarine Red W.B. . 2 ,,	

Another Method.

Bichromate of Potash . 7 lbs.
Tartar   .   . $3\frac{1}{2}$,,
Oxalic Acid   .   . $1\frac{3}{4}$,,

Dye.

Aliz. Blue D.N. 20
  per cent   .   25 lbs.
Aliz. Blue W.B.   . $1\frac{1}{2}$,,
Acetic Acid   . $1\frac{1}{2}$ gallons.

We merely give these as typical examples of a class of colours that can be varied almost indefinitely, but sufficient to show the change that has taken place during late years.

Note.—Students who wish to pursue the subject of dyeing further may consult such works as that of *Dyeing and Tissue Printing*, by W. Crookes, F.R.S., published by George Bell and Sons; *The Dyeing of*

Textile Fabrics, by J. J. Hummel, F.C.S. of the Yorkshire College, published by Cassell and Co. ; and the very recent *Manual of Dyeing*, in three large vols., by Knecht, Rawson, and Lowenthal, published by C. Griffin and Co., Limited. In these three works the subject of dyeing is exhaustively treated.

PART VIII

TEASING OR WILLEYING, BURRING, MIXING, AND OILING

THIS is the first operation after the wool is dyed, and is known by a variety of names, as teasing, willeying, willowing, and twilleying. There is but little doubt that the name is a corruption of the word "winnowing," as the operation is to shake up and blow out the dirt, dust, and refuse by means of a fan, with the addition of the modern name and operation of teazing, which means to pull to pieces and to open. The operation combines both the pulling of the matted wool to pieces and the blowing out of the dust, dirt, and loose dye-wares, etc. It is important that this operation, though a minor one, should be well performed, but most of the machines in use for this purpose are very defective. The operation is intended to loosen and open the matted wool and prepare it for the processes which follow. There are two objects in view in teazing or willeying; the first is to open the wool, expand it, and make it light and feathery, and the other is to blow out the dirt, to winnow or free it as far as practicable from all kinds of extraneous matter. The machine for accomplishing this is shown in Fig. 12, and section, Fig. 13 ; it is composed of a skeleton iron cylinder, upon which are fixed strong transverse bars of wood, into which are inserted double rows of spikes, as shown at C. The cylinder when working revolves at the rate of from four to five hundred

revolutions per minute, and the spikes or teeth work in
and between other rows of teeth fixed in the stationary
back rail B shown in the end section, and also in and between the teeth of the workers marked W W W, which are
rollers placed over the main cylinder, as shown in the end

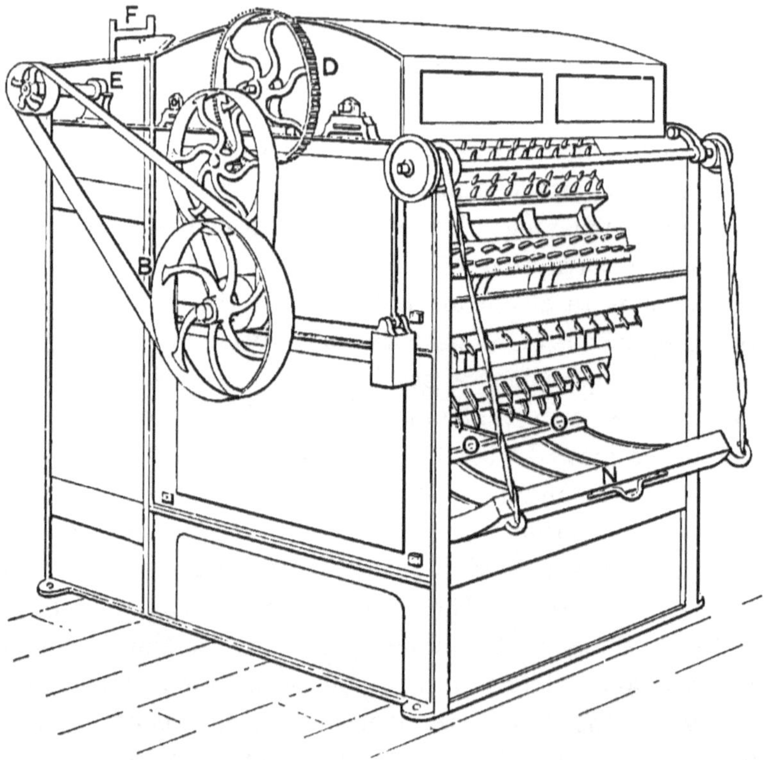

FIG. 12.—TEASER.

section, and in the box D of the sketch of the machine.
The want of the proper setting of these teeth to work in
and between each other constitutes one of the radical faults
of the majority of the machines at present in use.   A
glance at Fig. 12 at the portion marked C will show that
there are spaces between these spikes or teeth, and that

the teeth of the workers that are to work into the
cylinder teeth have to be so arranged as to stand in the
centre of the intervening spaces that occur between the
teeth of the main cylinder.   The workers W W W revolve
very slowly, but work into each other as well as into the
main cylinder, and the depth that the different portions
work into each other should not be less than from an inch

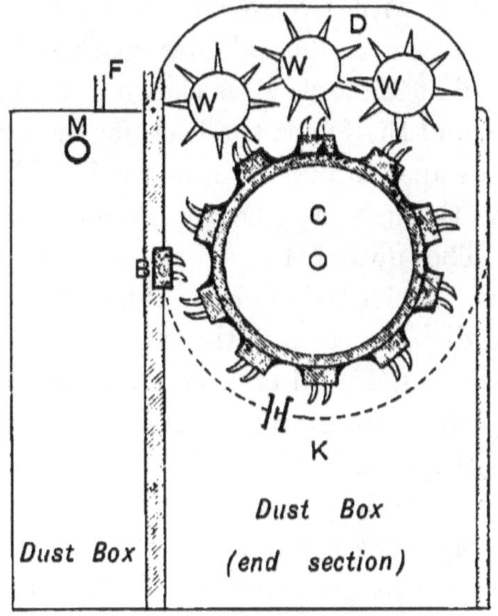

FIG. 13.—SECTION OF THE TEASER.

to an inch and a quarter.   The want of attention to this, or
the want of understanding the necessity of this, is the
radical fault in the setting and working of these machines.
Many machines are set with the teeth not more than a
quarter of an inch into each other, and the consequence
of this is that the wool only just catches on the points
and escapes again before being properly opened, so that
it becomes stringy by being twisted instead of opened.

The teeth not only need to be sufficiently set into each other, but they require to be accurately arranged so as to stand into the centre of each other's spaces, instead of having, as is too often the case, a wide space on one side of the tooth and only just a clearance on the other. This equal escapement space on each side of the tooth is requisite, otherwise the staple of the wool will be broken instead of being only opened. The machine consists of one cylinder with twelve spiked wings, three spiked rollers or workers over the cylinder marked W W W, and one spiked rail at the back, B, as shown in the end section of the machine, and the fan at the back for drawing away the dust, which escapes through the mouthpiece F, and is conveyed away through a pipe or chimney to a proper receptacle. The dotted line H is a wire grating that surrounds the lower portion of the cylinder, through which the dust and dirt escapes into the box K underneath. The workman places an armful of wool upon the let-down portion of the underneath grating at O O, and then closes the opening by lifting the handle at N, and thereby forces the wool against the rapidly revolving cylinder, with its frightful array of spiked teeth, running at the rate of four to five hundred revolutions per minute, and these force the wool through the teeth of the back rail B, and also through the spikes of the workers W W W, which meet the motion of the main cylinder. When the wool has been a sufficient time in the machine the opening door at N is again let down, the wool carefully removed, and a fresh quantity laid upon the grating and the operation repeated. This is a very dangerous machine to work, and at times they have been the cause of frightful accidents, when the men attending to them have happened to get caught by the revolving teeth, the material or wool having

to be put in and taken out in this dangerous manner
at the opening in the grating N, formed by a portion
of the grating being swivelled on one side and suspended
by cords and weights on the other, so as to pull down and
shut up again every time the wool was put in or taken
out. This unsafe manner of working the machine is now

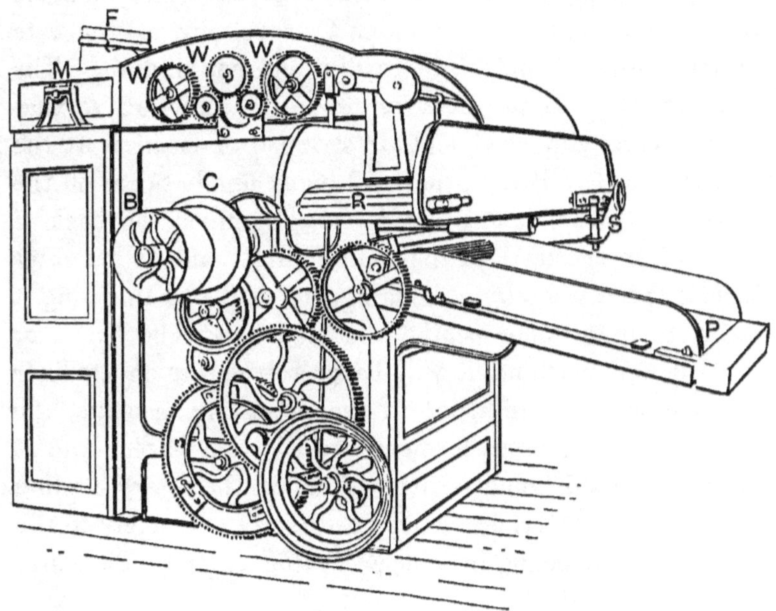

FIG. 14.—SELF-ACTING TEASER.

being done away with by the introduction of a self-acting
machine, shown at Fig. 14, which it is desirable should
come into general use. It feeds in the wool at intervals
from the apron or lattice P, and delivers it out of the
machine from the moving lattice R at an altitude sufficiently
high to be dropped straight into a sheet or bag hung up for
the purpose ; and by simply taking off and reversing the
cord S the wool can be delivered into the bag on the right

or left-hand side of the machine, as may be most desirable or convenient for the time being.

In carrying the preparation of the material a stage further, another machine is often employed called a "Fear-nought" or tenter-hook willey, represented in Fig. 15. The wool is fed on to the lattice A, and carried forward into the grip of the feed rollers at B, which delivers the wool to the large cylinder C, revolving at the rate of one hundred and sixty revolutions per minute. The workers D D D move in an opposite direction to that of the main cylinder C, and are stripped or cleaned by the stripping rollers E E E, the wool being finally taken off the machine by the doffing fan F working upon the back of the hook of the teeth of the main cylinder, and at a greater surface speed, thereby unhooking the wool and throwing it down upon the floor behind the machine. The hooks or teeth in this machine, it will be perceived, are much finer and closer set than those in the preceding machine, the teaser—or, as it was formerly called, the teazer—and it brings us unmistakably within sight of the rotary carding machine by its workers being set directly and solely to the main cylinder, being, in fact, a carding engine on a coarse preparatory scale.

Burring.—After the wool has been teased or opened by the teaser and fearnought, the next process is that of burring or moting in cases where the wool contains burrs or motes of any kind ; and now that our wool-supply is large and drawn from all parts of the world, many wools contain seeds of various kinds, either in the way of "burr," the round prickly sort being most obnoxious, the carroty seeds, or in some other form, such as being "*shivey,*" besides a great number of straight motes, very fine small sticks, bits of straw, etc.

To enable the manufacturer to cope with this class of wools a burring machine has become a necessity in every modern mill that puts forth any pretension to being abreast of the times in requisite appliances. There are few mills on the Continent without a Sykes's burring machine, and there certainly ought not to be a single mill in our own country without one of these handy and *very effective* machines. No preparatory set of machines is complete without it.

The bulk of the cheap South American wools find their

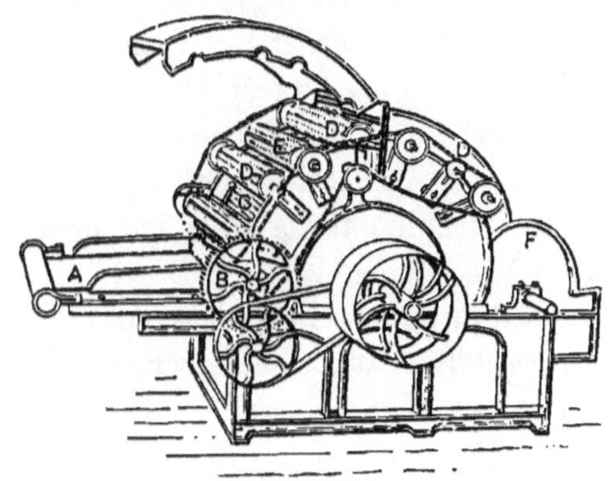

FIG. 15.—FEARNOUGHT.

way to the Continent. Only a small portion comparatively is consumed in the United Kingdom, chiefly on account of the "burr" which infests them, as well as the reluctance of many manufacturers to move out of the "rut" and equip themselves with suitable machinery to face any kind of wool that may tend to economy and efficiency of production. Notwithstanding all that has been said against the South American wools, there is a place for them (especially the Monte Videan and kindred wools) which they can fill to advantage in the fancy woollen

and other branches of our great wool industry, if manufacturers can only be induced to take the trouble to clean
them by equipping themselves with the requisite machinery
for scouring and burring them. The burring machine is a
marvel of ingenuity and is a lasting honour to its inventor,
as it enables wools to be used that prior to its invention
were comparatively useless, on account of their filthy
condition in regard to "burr" and other extraneous
matter. The burring machine and other mechanical
contrivances tending to the same end are the appliances
that ought to be used *instead of half destroying the wool* by
saturating it with acid in order to "extract" or get rid of
the vegetable matter in the way of "burr" and other kinds
of mote. *Mechanical* and *not chemical* are the means that
ought to be resorted to in dealing with dirty wools. In the
Sykes's burring machine the wool is fed upon the lattice
sheet A, Fig. 16, as shown in the section (Fig. 17), which
carries it forward into the grip of the small feed rollers B B,
and they in turn deliver it to the beater C, which throws it
upon a second lattice feed-sheet D. This second lattice
sheet D carries the wool under the large brush cylinder
E, and presses it into the interstices of the brush. This
large brush cylinder E then takes the wool in the direction
of the arrow until it comes into contact with the comb
cylinder F, which is a marvel of elaborate and ingenious
construction; it is covered transversely with steel comb
plates throughout its entire circumference from 1 inch
to $1\frac{1}{2}$ inches wide, out of which project fine needle-
pointed teeth on the side of the plate corresponding to
the direction of the arrow and to the direction in which
the cylinder revolves when at work. These fine needle
teeth point slightly towards the surface of the cylinder,
but do not stand above the general surface, and conse-

quently do not appear in the end section of the machine.
As the comb cylinder F revolves at a much quicker speed
than the brush cylinder E, its fine needle teeth take each a
few hairs of wool off the surface of the brush cylinder E in
sweeping over it.   The comb cylinder rapidly conveys the
thin film of wool forward until it meets the finely-adjusted
edge of the ledger blade G, the bevelled edge of which stands
so close to the cylinder that it refuses to allow anything to

FIG. 16.—SYKES'S BURRING MACHINE.

pass that is more bulky than a few hairs of wool in
thickness ; therefore any "burrs," seeds, or motes of any
kind are made to stand challenge at this point by the fine
bevel edge of the ledger blade, which runs transversely
across the entire cylinder, as shown in the section at G,
and against it the spiral cylinder H rapidly revolves in the
reverse direction to that of the comb cylinder, beating from
off the edge of the blade G any burrs, seeds, bits of straw,
sticks, shives, or motes of any kind that the ledger blade
has intercepted and casting them down upon the grating J,

from which they are speedily driven by the beater K into a proper receptacle. The wool in the fine needle teeth of the comb cylinder has passed the challenge of the ledger blade, by its action, and that of the spiral cylinder H, it has been stripped of its burrs, seeds, and other motes, and is borne onwards till it reaches the brush cylinder L, which revolves at greater speed than the comb cylinder and in the opposite direction, and consequently sweeps off the cleaned wool from the comb into the box M at the back part of the machine, and in this manner the operation continues, each portion of the machine contributing its quota towards the general result. The small roller N picks off any stray locks of wool that may be found still adhering to the large brush E after the comb has swept over it, and lays them upon the brush P, which offers them again to the comb cylinder; P in turn is served by the beater Q throwing them down as surplus, if any, upon the lattice D to take turn for a second round of duty with the fresh arriving wool.

Of course, the setting of the ledger blade has to be varied according to the kind of wool in hand; if the wool is long and coarse the blade and spiral must be set farther away from the comb cylinder, so as to allow more room for the coarser, bulkier material to pass through; and, on the contrary, when the wool is fine and short the setting must be proportionately finer to suit the material. This machine is no exception to the general run of machines; to be efficiently worked it must first be *thoroughly understood*. It has been charged against it that it breaks the staple of the wool; the writer's long experience of this machine does not support such a charge. Almost any machine can be set in such a manner as to damage the material that it is at work upon, but that is the fault of the person in charge and not the fault of the machine. In the earlier opening processes, whether teasing, burring, or carding, there is inevitably some slight breakage of the material, and allowing for argument's sake that this is the case with the burring machine, still there need be no more breakage than takes place in the ordinary working of a first or breaker-carding machine. Why do we call the

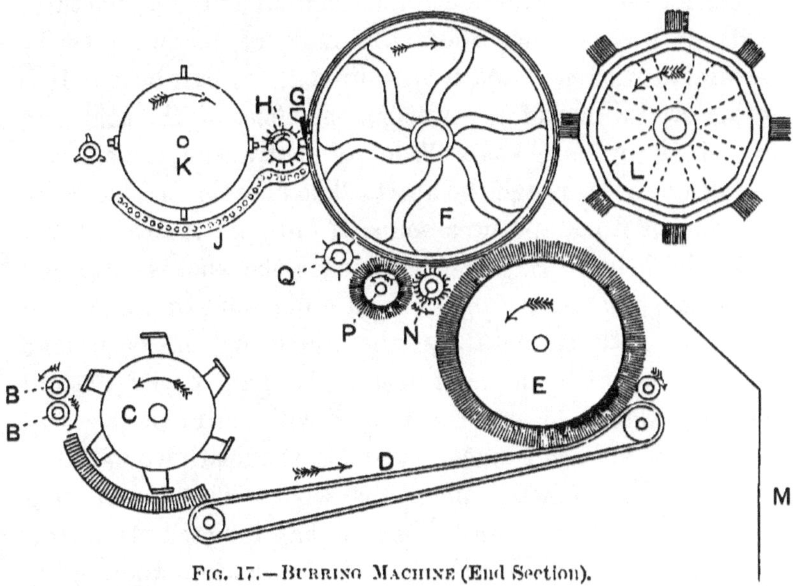

FIG. 17.—BURRING MACHINE (End Section).

first machine of a set of carding machinery the breaker? Because the first machine has to loosen and forcibly break open the matted locks of the wool, which the cleverest people are not clever enough to break open without some slight amount of breakage. Any amount of breakage that may result from using the burring machine is more than compensated for in the saving of cards effected in the forepart of the first carding machine or breaker in consequence of the thorough opening that the wool gets in passing through a burring machine when intelligently set. The saving in cards alone will very soon pay the cost of a burring machine. There is one rule that ought to be laid down in manufacturing, and adhered to on all occasions, *never to do chemically* anything that *can be done mechanically*. In wool sorting throw aside the most burry pieces to be extracted if preferred, but the thoughtful and intelligent manufacturer will on no account allow the bulk of his wool to be saturated with acid and then subjected to high heat. It is of no use talking of neutralising the effect of the acid in a soda bath; we might as well talk of neutralising the effect of firing a bullet through the heart. The mischief is done when the shot is fired, and no amount of surgical treatment will bring back life. So in the case when the wool is saturated with acid, *the mischief is done*, and no amount of immersion in a soda bath can neutralise the injury that has been done to the wool, or in the least restore it again to its natural healthy state. True, the soda bath will neutralise the acid, but the soda bath cannot neutralise the death-wound that the acid has inflicted upon the wool. The neutralising farce is not begun until the acid has inflicted its death-wound upon the wool. This is an instance where thoroughly grounded technical and scientific knowledge requires to be brought to bear upon the manufacturing processes—not half-grounded knowledge that only knows that acid will destroy vegetable matter, but does not know that it will seriously injure so highly organised a structure as the wool fibre. The wool fibre is a product of animal life, and is of infinitely higher and more sensitive organisation than that of any vegetable product, and much sooner injured. Instead of following half-informed science, the rational course to take is in sorting the wool to throw the most burry portion on one side to be dealt with separately, and put the bulk through a good burring machine, which, if properly set, will take out the "burr" without materially breaking the staple. The most burry parts thrown aside could be passed two or three times through the burring machine, and then, if not sufficiently clean, and if nothing else would satisfy, could be subjected to a weak extracting bath, as any motes left in it would be very small and not need strong treatment. If this course were followed there would be no need to subject the bulk to the harsh treatment of "extracting." The breakage of staple by the burring machine is not a tithe of what takes place in the carding of wool that has been extracted, and this, too, must be borne in mind, that wool that has gone through a burring machine does not suffer any further breakage worth mentioning in passing through the breaker part of a set of carding machines. A certain amount of breakage is inevitable in opening out the staple, and when the wool has passed through the burring machine the opening breakage has not to occur later on in the forepart of the set of carding engines. A burring machine must be studied before it can be understood and worked properly, but when understood is one of the most useful machines in a woollen manufactory and deserves better attention than it receives.

Mixing.—After the willeying and burring of the wool comes the mixing, or, as it is called in some districts, "blending." This operation is usually performed on the wide open floor of the willey room, and consists of the mixing together of two or more different coloured wools, or of two or more varieties of materials. Sometimes the mixture consists of different coloured wools, with flocks, shoddy, noils, mungo, or some of these along with some one or more of the almost endless varieties of prepared woollen material that pass under the name of "extract," "recovered wool," and so forth. Or further, again, the mixing may contain some part of the foregoing along with the addition of some vegetable fibre, such as cotton, China grass, etc., or some other kind of animal fibre, as camel hair noils, alpaca noils, or mohair. Silk noils and waste are often mixed with woollen material. Let no one be alarmed at the enumeration of such a list of what may be considered "vile" materials, and think that we are fallen upon nothing but degenerate days, and that there is no such thing as wool used in its purity in our times ; such is not the fact, as there is as much, or more, wool used in its purity as ever there was, and those who are still willing to pay the price for all-wool goods can have them, but human society is so graded as to "ways and means" that the manufacturer, in these days of keen competition, has to widen the area of his market by taking into account the purchasing power of all grades and classes. Those people who inveigh against the use of shoddy frequently do not know either what it is, or what is its use, and therefore do not display their wisdom, but the reverse. There are few things in these modern days that are not turned to some useful account, and it is nothing but prejudice, and that, too, of a very unreasonable kind, that objects to the utilisation of anything that can be turned to the good of mankind. Why object to the use of shoddy ? Why object to its being turned into cheap, useful, and warm clothing ? It would be just as reasonable to object to linen and cotton rags being turned into paper, or to vile-smelling gas tar being turned into brilliant aniline dyes, or to the utilisation of anything that has hitherto been considered a waste product. All such objections are simply foolish prejudices. Whether we are prejudiced against the use of shoddy or not, all must bow to the logic of facts, as we import something like 90,000,000 lbs. of rags, and altogether we use annually over 120,000,000 lbs. of rags torn up into shoddy and mungo, etc. Shoddy, strictly speaking, cannot be called raw material, seeing it is manufactured, though there is one sense in which it may be regarded as such—the manufacturer buys it as he buys his other materials, and in this sense it is raw material to him. Shoddy is worn-out garments of soft knitted, netted, and woven woollen that have not been felted or milled, such as hosiery, netted shawls, flannels, French merinoes, and soft Berlin wool work, etc., taken and torn thread from thread and fibre from fibre by a kindred class of machine to those we have been discussing in this chapter. The teaser tears and opens the locks of matted wool, fibre from fibre, and the rag machine, as it is called, tears and opens the knitted rag, thread from thread and fibre from fibre, and delivers it in loose flakes, after the manner in which the teaser delivers wool. Mungo after a like manner is produced from felted or " milled " cloth rags, and is consequently much shorter in length of staple or fibre on account of the felted or matted nature of the material from which it is produced.

Touching this matter of shortness in the staple of material mixed in along with wool in the woollen manufacture, it has been wittily said that anything can be used of the nature of woollen, providing it "has a fibre long enough to have two ends." This is certainly an extreme statement of the case, and must be taken with "a little salt" and freely discounted in practice. It is quite true, as we shall see farther on, that the woollen thread is of such a structure as to admit of comparatively very short material being used in the woollen manufacture, but the intelligent manufacturer will use a wise discretion in the mixing of his material, as in everything else, and will provide himself with a little of "middle," as regards length of fibre, in addition to the "two ends." The bounds are frequently overstepped at this point by grasping at too low a material, thinking thereby to accomplish the production of low-priced yarns and goods. It often happens that in descending too low in first cost, the resulting outcome or product proves to be very dear, and a much better result would have been obtained by mixing a better material at higher first cost that would have worked better and made much less waste. Waste is such a reducing item in the net result that it is necessary to limit the consideration of first cost by what can be got out of the material selected. Many a manufacturer has mixed together material of greater first cost than some of his neighbours, and yet with the higher-priced materials has brought out better and cheaper yarns and goods, in consequence of the material making a better weight and less waste. The tendency to low first cost must always be held in check by the prospective net result. It is not always found in practice that the lowest-priced materials bring out the lowest-priced goods, but very often the outcome is far otherwise. Except the material is brought up to be within range of its purpose in quality, staple, and general working capability, excess of waste and other reducing causes will more than counterbalance any small difference in first cost.

It is clear that there is a possibility of overstepping the legitimate bounds of cheapness in first cost of material through lack of intelligence to forecast the outcome or product of certain materials, so that in buying and judging material it is necessary to consider the producing capacity of the article before you ; whether its capacity of production will not more than cover any difference in cost, as compared with a certain other kindred article of a lower price. First cost in materials, in and of itself, must always be considered as relative ; capability of production only can make it absolute, and nothing but a thoughtful, careful selection and combination of materials, and adapting the beginnings to the calculated ends of the manufacture, can bring out uniform and satisfactory results. The person who has to mix and combine these various materials so as to meet the desired end and aim in the manufacture should also be the purchaser of the various materials, wherever practicable. It is a great advantage when buyer and user can be one and the same person ; divided responsibility at this particular point never works well. But whatever materials the manufacturer, in the free exercise of his discretion, may have purchased, they are now to be mixed together in such proportions as a sound discriminating judgment may determine, according to the class of goods that the manufacturer is aiming to produce. First a layer of the teased or willeyed wool a few inches in thickness is spread out very evenly upon the clean-swept teasing room floor in square form, then a layer of noils a few inches deep, exactly covering the layer of wool, then a layer of shoddy is spread over the same area, each layer being levelled by the use of long rods, and each layer regulated in thickness according to the respective quantities of each kind of material intended to be mixed together. Should the mixing or blend consist of the three materials named, then a second layer of wool, noils, and shoddy follow each other again, and the routine of succession is repeated until the whole of the material for the mixing is piled up on the square, which not unfrequently rises to a couple of yards in height, and extends over a great many yards in area.

The experienced workman can soon settle as to how much area he should cover with his mixing when he sees the gross quantity of material he has to mix. Take a second illustration of a mixing, say a layer of willeyed wool as before, thrown down in armfuls as it is taken out of the willey or teaser, then a layer of mungo, which is run through the teaser to shake it loose and to lighten it up, so as to enable it to mix more freely; then follows a second layer of wool, willeyed by armfuls as required to make the square pile, then again a second layer of mungo as before. If the mungo consists of two kinds, each kind is placed upon the pile in alternate layers, one kind being used to form one layer, and the other kind being used to form the next layer when the mungo comes in turn. The same applies when there is more than one colour or kind of wool; a layer of each kind in turn must be placed upon the pile as the turn for wool comes round. A certain rotation of layers has to be adopted at the commencement of a mixing, and the rotation is repeated until the whole of the material is piled together into a square heap. Of course, the various layers can only be equal in thickness when the various materials are in equal quantities, or are equal in bulk. Supposing there are 1000 lbs. weight of a certain colour of wool, and but 100 lbs. in weight of another kind; in such a case the layer of the lesser weight can only consist of a sprinkling, but if only a sprinkling, it must be very evenly spread over the entire area of the pile, or the mixing will be irregular, and the work will be spoiled. Sometimes the lesser quantities do not amount to more than a fiftieth part of the greater bulk, in which case it is usual to take say double the quantity of the smaller weight from off the bulk and mix this and the small weight together first before commencing to spread out

the bulk, and by this means the small quantity is increased
and made more manageable.   An illustration like the following will make the meaning more plain. Say we have 1000 lbs.
weight of black wool, and only 50 lbs. weight of either bright
orange or pure white, which we wish to mix together.   The
course we should adopt would be to take 100 lbs. out of the
bulk of black and form a small mixture with the 50 lbs. of
bright orange.   First make a small square layer 6 inches
thick of black, then a layer of bright orange 3 inches
thick, and so on alternately till the orange is all on the
pile, and treat it as a separate mixing until the two are
evenly united, then it is laid aside and considered as orange
for forming layers in mixing the bulk ; by resorting to this
method the orange is increased threefold in bulk, and can
be more evenly mixed.   A like mode is followed of proceeding by layers, even if the material consists of nothing
but pure wool, and that, too, of the finest quality ;
layer is spread evenly upon layer until the whole is got
together, there being unevenness in dye, drying, and other
things, combined with great differences in spinning properties in different portions of the wool, so that the same
care of equalisation and mixing is necessary in the laying
together all-wool material as in putting together material
of an inferior character.

It is a great mistake to suppose that an all-wool mixing
does not require as much care as any other; its greater
value alone should entitle it to equal consideration.   Too
great attention cannot be bestowed in order to get the mixing even, as an inequality cannot afterwards be put right,
the unevenness remaining throughout the entire series of
operations as a permanent defect.   In this way the various
materials, however many in number and variety, are layered
one upon another, and as uniform in thickness as is prac-

tically possible, keeping the pile just as wide at the top as it is at the bottom.

How to treat the huge pile so formed is the next point for consideration. We see how carefully the square pile has been formed, layer after layer, from the bottom upwards. Now a somewhat different method is taken, still having the same object and end in view—evenness and regularity of the mixing. The pile being formed, the next move is to commence pulling in single handfuls all along one side, carefully taking a little from each layer from top to bottom in breadths of about a yard wide. This is continued till the whole of one side of the square has been gone through, always pulling the whole of the side equally, so that when the end is reached the whole side presents a perpendicular and even frontage. That which has in this way been pulled from the pile is tossed about and mixed together by means of short sticks or rods, then run through the teaser to mix it more thoroughly, and then spread out evenly to form the base layer of another pile of the same size. This done, another portion has to be similarly pulled down, care always being taken to pull down evenly from top to bottom; mix well together, run through the teaser as before, and throw it evenly over the previous layer of the new pile. These operations are repeated till the whole of the first pile is pulled down and transferred to the second. By this double series of layers well managed the material can be evenly laid together, and the mixing is completed by treating the second pile as the first was treated, by taking it down from off one side and again running it through the teaser or willey. When any cotton, silk, or other fibre than wool has to be added to the mixing, it can only be safely done after the oil has been applied and absorbed by the woollen materials, for if the oil comes directly into contact with either silk, cotton, or China grass, it prevents the cotton, etc., from opening freely in the carding process. Therefore when silk, cotton, China grass, or other extraneous fibre is required to be introduced, it can only be done safely and effectually by commencing the series of layers again, after the woollen material has been oiled and the oil absorbed, and inserting in these layers the cotton, silk, China grass, or whatever else the extra fibre may be, going through the whole round of mixing as in the first instance.

Oiling.—A great deal of ingenuity has been bestowed upon the invention of machinery for oiling wool, but the operation is so simple, and can be so easily and quickly accomplished during the mixing, that it seems a pity to waste capital and ingenuity upon machinery for such a purpose. While the material is spread out in layers for mixing, a common-sized watering-can, to which is attached a T-shaped rose, is taken and filled with oil, the oil evenly sprinkled over the entire area of the layer, say of wool, and this simple operation is gone through every time a fresh layer of wool is laid down. Should the mixing consist of a variety of materials, then some one of the materials should be fixed upon, and the oil should be sprinkled upon it every time it comes round in the layer. For instance, in some cases it is desirable that the oil should be spread upon the layers of wool, in which case the oil is spread over the pile every time a layer of wool is laid down. At another time it may be advisable to spread the oil upon the shoddy layers, especially if the shoddy has been dyed just previous to mixing. This is a matter of detail, and the course most desirable to take can only be settled by sound practical judgment on the spot and at the time. There is a great error that is often fallen into which should be pointed out, and that is not getting the wool or other material thoroughly dried. Unless the wool, etc., be thoroughly dried, the dust cannot be got out in the teasing, and if the dust be left in, then when the oil is applied, and the mixing taken to the card room, a paste forms by means of the dust and oil, which speedily clogs up the mouth of the cards, and produces waste and great irregularity in the work. By all means let the wool be thoroughly dried, but not scorched, before teasing, so as to completely get rid of the dust, at least as far as practicable; for not only does the dust, if left in, form a paste by combining with the oil, but it to that extent robs the wool of the benefit of the oil for carding purposes. The dust being got rid of, the wool can be damped back again to any degree that may be deemed advisable to facilitate its working. In a large manufactory there is always a quantity of condensed steam-water to be obtained, and a long number of years' experience has taught the writer that condensed steam-water is the best agent to employ along with the oil for conditioning the wool for working. To ten gallons of oil take five gallons of condensed steam-water, into which has been poured half a pint of ammonia, and stir the water and the oil together. The small quantity of ammonia will enable the soft steam-water to unite with the oil and form a very limpid oil cream, which is not so viscid, and is much less dense than pure oil alone, and works much better. If too much ammonia is used the compound becomes too stiff, if too little, the water and the oil do not sufficiently and evenly unite; judgment is required so as to have just sufficient ammonia in the water to hold the oil in solution long enough to give time for its being spread upon the wool. This caution is necessary, as ammonia varies considerably in strength, and the quantity used must be varied accordingly. There have been a great many nostrums brought out during the last forty years as substitutes for oil, but I would recommend that nothing but good olive oil be used. Where the work is of an inferior kind, and not worth the bestowal of good oil, it is still advisable to use good oil, but less in quantity according to the inferiority of the work, and thin it out with steam-water as above indicated. Mungo and shoddy are often spoiled by being pulled in black and other rubbishy kinds of cheap recovered oils. Olive oil, in and of itself, is perhaps a little too viscid, but thinned down with soft steam-water in the manner above described, spreads freely, and diffuses itself over the entire mass of a woollen mixing, so that one portion is not saturated while other portions are left dry, but all portions participate alike. The proportion of two of oil to one of water is given as being applicable to all kinds of better-class work; any other proportion can be adopted so as to suit all kinds of work. Ammonia being a volatile alkali flies off as the work passes through the card room, and leaves no ill effects behind. Indeed, the cards work much cleaner with the oil thinned out, as every portion of the material comes into contact with it, than when pure oil alone is used, in consequence of the pure oil being a little too viscid. As the result of these remarks no one must conclude that it is possible to do with one-third less oil. The object is to point out that by adopting this suggestion one portion of the mixing does not get more than its share of oil while the other is left without. The oil is spread or thinned out, so that every portion of the mixing is brought under the influence of its effects, and the mixing cards freely, and with much less waste. The quantity of oil to be applied, say to every 100 lbs. of wool or other woollen material, varies with every individual class of the wool industry, so that no rule can be laid down. The best middle-class manufacturer often applies two gallons of oil to every 100 lbs. clean weight of material, while in the west of England more is applied. The other woollen industries make use of less. Each uses oil according to its own special needs. The oil, being applied while the material is being put together in layers, is equally diffused throughout the entire mass in a more reliable and satisfactory manner than can be done by machinery after the mixing is completed.

PART IX

CARDING

UP to this point the wool to some extent has been opened by the preparatory process through which it has gone, but the carding is especially and essentially an opening process, and that, too, in a most minute degree. Carding for the woollen manufacture is to some extent different from that of carding for any other textile manufacture, and proceeds upon somewhat different lines. Carding for every other industry except woollen inclines towards the combing principle, and to a greater or lesser degree aims at laying the fibres parallel. Woollen carding has no such object; its only cardinal aim is to open the material fibre from fibre, and to present it in a perfectly loose and thoroughly open condition, without leaning towards any artificial arrangement, to open out any and every natural arrangement of fibres that may have taken place during the growth of the wool, and to throw every kind of order into a chaos of confusion as a preparation for a fresh and very peculiar artificial arrangement, special to the woollen, and that does not pertain in anything like the same degree to any other textile manufacture. The object of woollen carding then, it will be clearly understood, is to break up all previous artificial or natural arrangement of the fibres of the material, and to present every fibre disengaged and disentangled, thoroughly opened out and free. In no other textile industry is the carding of the same importance and required to be done so uniformly as it is in woollen, and in no other industry has any irregularity such disastrous consequences. In the woollen processes there are no points or places at which a carding fault can be redeemed or atoned for. In every other textile manufacture there are these points or passing places where any inequality can be distributed or equalised, but the processes of the woollen afford no such opportunities; a point missed or a point neglected cannot be afterwards redeemed, and this will be more strikingly apparent as our discussion and illustration proceeds. Granting, then, that the woollen carding is of such importance in the manufacture to which it belongs, it is very essential that it should be thoroughly understood. To obtain a thorough insight into it, it is necessary to go back to some of the primitive modes of carding, for the principles can in many instances be more easily understood in some of the old hand processes than in the complex modern machines. It would be next to impossible to explain to any one the principle of carding by taking them into one of our modern mills and showing them the huge woollen carding machines; they would be bewildered by the multiplicity of the parts and the number and variety of movements. Let us turn aside from the big machines and endeavour to master the elements of carding as it was performed in its simplest manner in our domestic manufacture one hundred years ago, and then we will endeavour to pick out the corresponding parts and movements in the big machines behind us. The old domestic carding was done on two pieces of wood covered with card-cloth, each about 11 to 12 inches long and 5 inches broad, and it is

Fig. 18.—Hand Cards.

from the 5-inch breadth of the old hand card that we derive our present standard card-width, on which are based all our card numbers and card counts. These card-boards had each a short handle attached to the back of the board, as shown in Fig. 18.

This small oblong board was covered with card-cloth, which is composed of thin leather filled with points of fine wire, as shown in Fig. 18, and the wool to be carded was spread upon it; when filled it was placed upon the thigh of the workman, with the card teeth or points upwards, and with the handle pointing from the operator, and was held there with the left hand by means of the handle. The other card-board was held by the right hand, with the teeth downward, and in an opposite direction to that of the teeth of the left-hand card, and was then drawn gently over the left-hand card, which was held still and firm against the thigh of the workman while all the movement was effected by the right-hand card. This operation of drawing one card over the other by the right hand was repeated until the operator thought the wool was completely torn open and blended by the action of the opposing teeth, then, when sufficiently carded, the left-hand card was taken up from its position and the handles of the two card-boards were brought together into a position pointing towards the body of the operator, and with the teeth of both pointing in the same direction, when there was commenced a peculiar shuffling movement of the card-boards one against the other, which had the effect of collecting and bringing off the wool in a roll, or "roveling," leaving both top and bottom cards completely empty and ready to commence the operation anew. The roveling or roll of carded wool was now ready for the one-thread spinning wheel, as shown in Fig. 51 ; the one-thread wheel was capable of turning off seven times more work than the distaff, which was a great advance.

These small hand card jack-boards were followed by the carding-stock, or, as it was called in some districts, the tumming-stock ; tumming and scribbling being synonymous terms, both meaning the early part of the carding process. The tumming or carding - stock was slightly different in form in different districts. In some districts the carding-stock was simply a trestle or long topped stool, upon one end of which a piece of card-cloth was nailed down as a fixture, the remaining portion of the stool forming a seat for the operator or workman, and the fixed or fast-nailed-down card was double or treble the size of the small loose hand card-boards, so that twice the amount of work could be done by one person to what could be done previously by the small loose hand card-boards. Mr. Jubb in his *History of the Shoddy Trade* describes a carding-stock that is a little more elaborate in construction. He says, firstly, a frame composed of a seat for the worker with an inclined plane, somewhat in the form of a desk, before it ; secondly, square boards set with cards, having handles affixed ; on the face of the inclined plane or sloping board were fixed iron or steel teeth or coarse cards, upon which was placed the wool ; the operative (a word, we believe, not then applied in this sense) proceeded to draw the loose hand cards over the fixed ones with a kind of horizontal or see-saw motion of the arms to and fro.

The process of carding on the carding-stock was further

facilitated by suspending the loose or upper card by a pulley and cord from the ceiling, with a weight to balance it, so that the workman or workwoman had only to move the upper card backward and forward by means of the two handles that stood up from the back of the card-board. The next move onwards from the carding-stock was the rotary carding machine or engine, which has developed into the ponderous and elaborate machine of the present day.

But at this point we require to attempt that most difficult feat of observing something that is under our noses. Rousseau remarks that "You require much philosophy to observe accurately things which are under your noses," and upon our success in observing and mastering the simple elementary movements of hand carding will depend our ability to dissect the movements of the large machine, though we may be ever so familiar with the sight of it in the rounds of our daily duties; seeing is not observing. All persons that are in any way acquainted with mill life are quite familiar with the carding process, and are therefore in great danger of being content with something more superficial than an accurate well-grounded elementary knowledge of the foundation principles that underlie the woollen carding process, upon an accurate knowledge of which depends all future improvement and progress in our study of the carding department of manufacture. With this purpose in view, and before we try to master the movements of our large rotary carding machines, let us try our "prentice hands" upon its lesser and more humble relatives and predecessors, the hand jack-cards and the hand carding-stock. In using the small jack-cards, as well as in using the carding-stock, the wool had first to be spread upon the bottom card, and the loose or upper card pulled gently over the surface of the under or stock card several times, tooth working against tooth, until the wool was sufficiently carded or opened ; it had then to be got out of the cards, both top and bottom, to make room for a fresh quantity to be carded in the same manner. It is here necessary to observe that the first part of the carding process is to hook the raw locks of wool upon the innumerable card points of the bottom card in order to tear open and part them asunder fibre from fibre, till the whole surface of both top and bottom cards is covered with individual or single fibres, each one hooked upon some one of the separate card points. The other part of the process is that when the locks or tufts of wool have been thoroughly opened and separated fibre from fibre, by being caught and impaled upon some one or other of the card points or hooks, another operation has to be commenced, that of unhooking (or, as it is technically termed, "stripping") all the fibres of wool out of the cards, completely emptying them, so as to make them ready for repeating the operation afresh. Observe further, that as the first part of the carding process is hooking in order to open the wool, so the second part of the process is unhooking in order to clear it away, the second part being exactly the converse or reverse of the first, and in this manner the whole process is a perpetual repetition of two alternating motions. How this alternation of motions is accomplished in the continuously moving machine we shall discover presently, providing we have rightly learned our lesson from the hand cards ; suffice it to say just now that in the manipulation of the hand cards both these operations of hooking and unhooking or *carding* and *stripping* can be accomplished by one and the same card, by simply reversing its motion, one card performing the duty of two offices, but in the continuously rotating machine this cannot conveniently be done, so that a division of labour, the characteristic of a machine age, is resorted to, and the working or hooking card has to be provided with its attendant stripping or unhooking card, seeing that the working cards are precluded in the machine from stripping themselves, in consequence of their continuous motion in one direction. *Carding* and *stripping* or *hooking* and *unhooking*, be it understood, are the two distinct sections of movement that go to constitute the entire carding process, and in the machine, when rightly constructed and applied, adequate provision has to be made for each, and adequate provision is made in most machines, but not always rightly applied. Let us now endeavour to dissect the modern carding machine, and try to find out the duplicates of those simple hand card movements which we have been studying, *carding* and *stripping* or *hooking* and *unhooking*. The carding card is often called the working card, in order to distinguish it from the stripping card. Fig. 19 is a section (an end section) of the working parts of a modern carding machine. The large cylinder C takes the place of the bottom card in the hand carding-boards, and it also takes the place of the fast-nailed-down bottom card of the hand carding-stock; the rollers marked W W W W W are so many upper hand cards in the hand-carding process, and perform the same duty, only as the large cylinder C has so much surface, there is ample room for five of the upper cards to be at work at the same time, without being in one another's way, or in any way interfering with each other.

These five rollers, each marked W, are exact duplicates of each other, and perform exactly duplicate duties, so that when the student comes to understand thoroughly the relationship between the big bottom card C and the little

top card W and its connections, a good elementary foundation will have been laid on which to build a knowledge
of carding. In the hand-carding the wool was spread
upon the bottom card, and the top card was moved
gently over the surface of the bottom card, and this is
precisely what is done on the machine. The wool is spread
upon the big bottom cylinder card at R, and the little
top card W, called a "worker," moves slowly over the

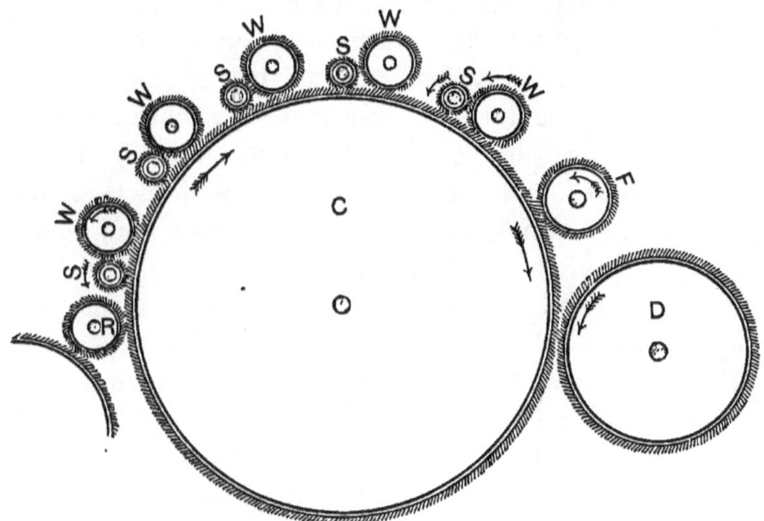

Fig. 19.—End Section, Carding Machine.

surface of the big bottom card C, the hook points of the
teeth of the cards meeting and passing each with the teeth
pointing in opposite directions, or face to face, as shown in
the section above, and by that means the wool gets carded,
or opened; each part of the machine moves in the direction
of the arrows. The big bottom card C moves at a much
quicker speed than the upper card W; but it is the cards
passing each other, with the teeth pointing in opposite
directions, or *face to face*, that cards the wool, and not their

relative speeds. The wool is presented to the large bottom cylinder card at R, and the cylinder C, moving quickly, seizes the wool with avidity, and bears it on in the direction indicated by the arrow; but it does not move far before it comes into contact with the little top card W, and both are speedily filled with carded wool. In the working of the hand cards we noticed that when the cards had become filled with carded wool the working motion had to be reversed, and a shuffling movement combined with it in order to empty out the carded wool; but in the machine this cannot be conveniently done. In the machine one member does one duty, but does that duty continuously so long as the machine is at work; so that the big bottom cylinder card C and the little top card W do nothing but card the wool, filling themselves continuously with carded wool whenever the machine is at work. The big card C and the little card W are called worker cards, because they *card* or *open* the wool; and as in the machine one part performs one duty only, we have now reached a point where there is another duty to be performed, and we must therefore look out another member of the machine to perform it. We have watched C and W fill themselves with carded wool, and we must now point out the two attendants that are standing at their masters' elbows awaiting the word of command to assist them to undress, and in the carding world this undressing is vulgarly and bluntly called "stripping." It will facilitate matters very much if we can succeed in constantly bearing in mind that the carding operation is a twofold process, consisting of "carding" and "stripping," or, to illustrate it in more popular language, "dressing" and "undressing." Every card that can be seen in motion throughout the whole length of one of these ponderous machines is busily engaged in performing either one or the other part of this twofold process.

The process is not half so complicated as it seems if we can only grasp hold of its two elementary motions, *working* and *stripping*.   Every member of this huge machine is employed upon either one half or the other half of this twofold process, so that we have not to puzzle our brains with a score of things, only two; and surely with patience and perseverance we can manage to grasp the elementary principles of two things, more especially considering that those *two* are only *two halves* of one process.   Every card that we can place our hands upon, from one end to the other of this long machine, is either a *working card* or a *stripping card*. Every working card has an attendant or stripping card, and as we have seen C and W fill themselves with carded wool, we must now hunt out the two attendant or stripping cards, to empty out, take off, or *strip*, and clear away the carded wool from each of the two working cards C and W.   Just above R in the end section is to be seen a little insignificant-looking card roller marked S, and this little roller is called a "stripper," and its duty is to strip the carded wool from off the larger roller close by marked W; it accomplishes this by being set closely against the worker W so that the friction of the one against the other can just be heard by an attentive listener.   The hook side of the point of the stripper card S rubs against the back of the card point of the worker W, and by this means strips off the wool from W in consequence of the face of S's card being set to work against the back of the card of W, and the greater speed at which S revolves.   Look at the section and examine very carefully the form and direction in which the teeth are bent, and the direction in which the teeth of the cards move, as indicated by the arrows.   When the card teeth meet each other face to face, or point against point, the cards in so meeting card the wool, and fill each other full of carded wool; but when the face of one card rubs against the back of the point of another, then the one that gets rubbed on the back by another running at a quicker speed is compelled to yield up its wool to its faster neighbour, and this is just what the stripper S compels the worker W to do—yield up its wool. The worker W moves on empty on one side until it comes into contact again with the big bottom card and gets re-filled; and constantly and continuously as its emptied portion gets refilled on one side, just as constantly and continuously is the active little stripper busily engaged in emptying it of the wool on the opposite side, passing it back again to the big cylinder in order that it may be carried forward to the next worker and stripper to undergo a like manipulation, and when these have contributed their share of duty the big cylinder carries the wool to the third, fourth, and fifth pairs of strippers and workers in rotation. Thus we see that as each worker (W) is continually being filled with wool on one side, and continually being emptied on the other, each stripper in turn delivers the wool back again to the main large bottom cylinder to be carried farther forward.

We have seen how each of the five sets of rollers marked S, W work and get quit of the worked wool; let us now try to understand how the big cylinder C is emptied or stripped of the wool. This feat is not quite so easily effected as in the case of the workers. The large cylinder C in facing the opposition of each of the five workers (W) in succession has the wool to some extent forced down into the spaces between the teeth of the cards, so that extra means have to be resorted to in order to clear out or strip the carded wool. The first part of the process towards accomplishing this operation falls to be the duty of the roller F, which is technically called the "fancy." To understand how it succeeds in this duty we shall require to notice very carefully the section, and observe the direction in which the teeth point in the card covering of the "fancy" compared with the direction of the card points on the cylinder, and the direction in which the cylinder moves when at work, and then we must be told that the surface-speed of the "fancy" is greater than that of any other of the revolving fraternity before us. We observe in looking at the section that F presents *not the face* but the *back* of its card points to the main cylinder C, and that those back points of F fall upon the back points of C, and that being the case, there are then no hook points presented direct by either; and as F has a greater surface-speed than C, it follows that F can act only as a wire brush. This is just what it does; it brushes the wool out of the large card cylinder C, and would send the wool flying all over the room were it not that its speed is moderated to such a nicety that it shall only be able to lift out the wool to the surface of C, which allows the wool to travel forwards on the surface of the cylinder till it meets the teeth on the cylinder D, which hook it off as C goes rapidly past on its way towards R, where it takes on a fresh supply of wool, and deals out a portion to each of the five pairs of workers and strippers, as in the previous round; and so the carding operation goes continuously on.

We have seen that D finally gets hold of the carded wool, and yields it up to a steel comb that takes it off at the opposite side of the doffer cylinder D to that on which it was received. The student will not act wisely if he seeks to hurry through what may seem to him a tedious detail. He should pause a considerable time at this point, and make a close study of the section of the carding machine (see Fig. 19), noting the direction of the arrows, and the direction

in which the teeth of the cards point, and read again and
again the letterpress description of their movements, till
each movement and its object become thoroughly and
ineffaceably impressed upon his mind. Only master these
two elementary movements fully, and the relative position
and direction of the points of the card teeth in each movement, and a good foundation will be laid for what is to
follow. Divide and conquer is a practical rule of life, and
it is in no case more true than in the present instance.
Dissect the carding machine into parts, and there is a
reasonable probability of understanding it. Take one
portion, and master it, and then it will be soon enough to
go on to the next point. The initial point to be mastered
in this instance is the question of the worker and the
stripper, and their relationship to the large cylinder; and
when we have mastered the movement and object of the
first pair of workers and strippers in our machines, we
have the key to the whole of the machine, and, in fact,
to the whole of the carding process. We begin with the first
pair of workers and strippers immediately over the letter R
in our section, and it is a question of worker and stripper
all round the circle of the large cylinder till we reach the
doffer, and the same thing repeated from beginning to end
of each machine. The whole carding process is composed of
a perpetual repetition of two alternating motions, working
and stripping, till the wool has become thoroughly opened
by being caught and impaled upon some one or other of
the thousands upon thousands of card points. The worker
roller moves slowly, and though its teeth point in the
direction facing the teeth of the main cylinder it moves
in the opposite direction, as will be seen by the arrows,
fighting as it retreats, and still keeping its face towards
the enemy; *and it is this face to face to each other at the*

point of passing that constitutes carding, and *the passing each other face to back that constitutes stripping*, and the passing *back to back* that constitutes *the action of the "fancy"* or brush. The worker roller moves slowly, and as it comes round, the wool is stripped off by the action of the stripping roller, which rubs against the back portion of the teeth, and as its speed is quicker than the worker, the wool is by this means unhooked (or, as it is technically called, "stripped") and carried down to the main cylinder again, from whence it came ; in this case the main cylinder takes the wool off the stripper by its greater speed and favourable inclination of point. The surface of the stripping card always moves at a greater velocity than the worker card that is being stripped, otherwise no stripping could be effected. The small stripper roller revolves at a speed midway between the slow speed of the worker and the quick speed of the main cylinder, and thereby strips off the wool from the worker, and is in turn stripped by the quicker speed of the main cylinder working against the back of the point of the stripper. And you will notice also that the stripper operates against the backs of the teeth of the worker roller, thereby unhooking the semi-carded wool, and carrying it round and down to the surface of the main cylinder, where it is quickly caught hold of again and carried forward, some of it to be again caught by the worker, while the other portions that have got further carded pass on to the next worker and stripper, to be caught again by the worker, and stripped again by its companion stripper. So the operation is repeated at every worker and stripper all round the machine, and from one machine to another, until every fibre is separated from its fellows, every lock of wool thoroughly opened, and the fibres thoroughly and evenly intermixed ; for it is an important part of modern carding to mix the different fibres of the locks or staples of wool grown upon different parts of the sheep's body, and a great number of sheep of different ages, all possessing different spinning capabilities. Hence arises the necessity of having the fibres from all the different parts of the body, and from different flocks of sheep, well intermixed, as well as those from the fleeces of different sheep mixed well with each other, as one sheep's wool differs from another very considerably in the same flock in its spinning property, when there is no difference in quality. In addition to every working card having its own stripping card, every card has its own sharpening card, and so on always in couples throughout the machine. When a carding machine is set, that is, when the parts are adjusted to their proper positions in the English manner of setting, each stripping card sharpens the points of its own worker card, and each fancy sharpens the points of its own cylinder card, so that the daily action of the machine when at work keeps itself in good working point and order. On the Continent the carding engineer sets his cards differently, and is under the necessity of grinding up his points almost weekly, causing much needless waste of wire, and variation in the keenness of the points from day to day; and further than this, the ground point is not nearly so efficient as the needle point obtained by the friction of card rubbing against card in their ordinary working, and the greater economy through wearing down the wire more slowly. Then again, as to relative working power of the different parts of the machine, the large cylinder stands its part against all the workers arrayed round its circle, and is itself the most powerful worker of them all; indeed, it has more working capacity than the whole of the workers combined, and presents many more teeth at each point of contact with the workers on account of its greater circumference or flatness of surface. We have now seen that the carding process consists of two simple operations, *working* and *stripping*, continuously repeated, not only from the beginning to the end of one machine, but from the beginning to the end of the entire set of machines, however large ; hence the necessity of studying thoroughly the action of one single pair of workers and strippers, and grasping the principle of carding in its simplest form as a key to the remainder. The action of one worker and stripper is essentially the action of all workers

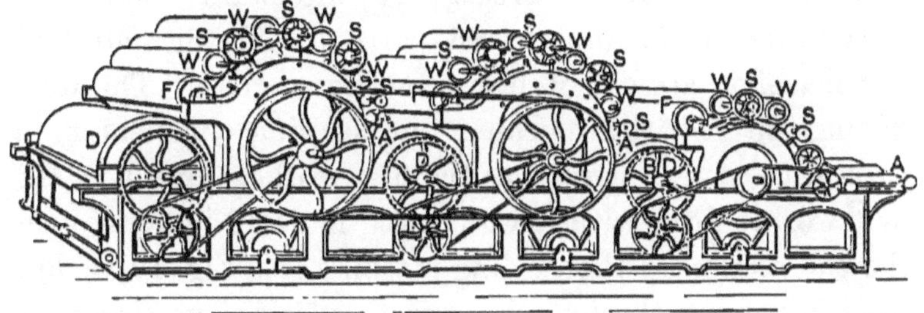

FIG. 20.—FIRST CARDING MACHINE, OR BREAKER SCRIBBLER.

and strippers, so that it is much better in every way to concentrate the attention upon the root idea of carding, as embodied in the worker and stripper in connection with the large cylinder representing the bottom card of the hand cards, than to let it wander at random over the entire machine before the initial principle has been grasped. Looked at as a whole, there is scarcely anything more bewildering than a large set of carding machines at work, with the wool moving from part to part, and tumbling over here and under there, as though the whole machine with its bustle of movement was instinct with life. Fig. 20 outlines a carding machine, the first of a set ; the wool is placed upon a lattice feed-sheet A, which is divided into sections, so that a certain given weight may be spread evenly over each section, one by one, as they come round in rotation. The lattice conveys the wool into the feed-rollers, and these feed - rollers gradually wind in the locks or staples of wool; the next roller is termed the "licker-in," and it carries the locks from the feed-rollers to the main cylinder, which, by the direction of the card teeth, will take the wool on till it comes into contact with the first worker, when the carding of the wool properly commences, as was observed in studying the end section of our carding machine. The first part of a first machine is called a breast-part; afterwards there is no breastwork to do, and breast-parts to intermediate and last machines are out of place, and do no good. The wool having entered the machine, the carding commences in earnest as soon as it reaches the first worker; and how the first worker deals with it we noticed in detail in going over the end section of the carding machine (see Fig. 19). The wool in passing through the breast undergoes precisely the same kind of manipulation as we indicated in our remarks on the end section, only in a modified form, but still the same essentially and in principle. In the breast-part of a carding machine the card clothing is not so fine, nor are the working parts set so close to each other, the object of breast-work being to open the wool a little, and to do it gently, so as not to break the staple, all the movements being slower than those which follow farther on in the carding process. There are two pairs of workers and strippers over the breast cylinder, and then comes the fancy, to get the wool out, or to brush it up to the surface of the breast cylinder card, so that it can be taken off by the breast doffer (B D). The breast doffer, having taken the wool off, acts the part of a large worker, and takes the wool round to its attendant stripper, which in this case is called the angle-stripper, and which works down in the angle formed by the breast doffer and the main cylinder of the next succeeding part of the machine, which is composed of the main cylinder and a full complement or series of pairs of workers and strippers, as arranged on the end section of the carding machine (see Fig. 19). This angle-stripper presents its face points to the back points of the breast doffer, and as it is running at a little quicker surface-speed it is enabled to strip the wool from the doffer in precisely the same manner as an ordinary stripper strips the wool from an ordinary worker and conveys it forward to the main cylinder card of the next part of the machine by presenting the back of its points to the face of the points of the main cylinder card ; and the cylinder, working at a greater surface-speed than the angle-stripper, takes off the wool, just as it would from an ordinary stripper. Briefly summing up, the angle-stripper working at a greater speed than the breast doffer (B D), and the large cylinder working at a greater surface-speed than the angle-stripper, by these means the wool passes from one to another, exactly on the principle that we saw exemplified in our study of the end section of the carding machine, always worker and stripper, continuously all through the machine, and all through the set of machines. Whatever technical names the parts are called by, the action is that of worker and stripper continuously repeated —the action of the top and bottom hand card in the old carding-stock over and over again on a large scale, and by a number of different names at different parts of the route, as the wool passes along ; every roller and every cylinder is either carding or stripping the wool, every worker takes the place of a top hand card, and the large cylinder takes the place of the bottom card to all of its workers.

The wool, having now passed through the breast-part and received a little gentle opening and reached the large main cylinder of the second part of the machine, takes exactly the same course from one pair of workers and strippers to another as was indicated in going through our study of the section (Fig. 19). The angle-stripper stands in the position of R in the section, and brings the wool forward from the breast-part of the machine and presents it to the large main cylinder, which carries it on till it comes into contact with the first, second, third, fourth, and fifth pairs of workers and strippers in succession, and thereby advances the carding a very considerable stage, till the wool reaches the fancy and the doffer as before, and is then taken off and transferred by the angle-stripper to the next main cylinder of the next succeeding part of the machine, to undergo the same kind of operation and routine as before. Finally reaching the last doffer, it is taken off by a steel doffing comb and conveyed by a Scotch or some other kind of self-acting feed to the intermediate machine of the set, on which the carding process has to be carried a stage further as the wool passes from part to part, until another self-acting feed places it upon the last machine of the set, which now mostly consists of only one cylinder, over which are arranged five or six pairs of workers and strippers with fancy and doffer, to which is attached the condenser.

The main cylinder in every instance, in every section of a set of machines, has the lion's share of the work to do; in fact, it does more work than all the other members put together; it not only does duty as a bottom working card against every top worker card in succession, but it does duty as general carrier, and conveys the wool from one worker to another, and does the further additional duty of stripping every stripper.

In our elementary study of carding upon the hand cards we saw that two working cards were always necessary, a top card and a bottom card, and that these two working cards had always to pass each other with the points face to face, or there could have been no carding of the wool; therefore, the main cylinder in machine-carding has to do

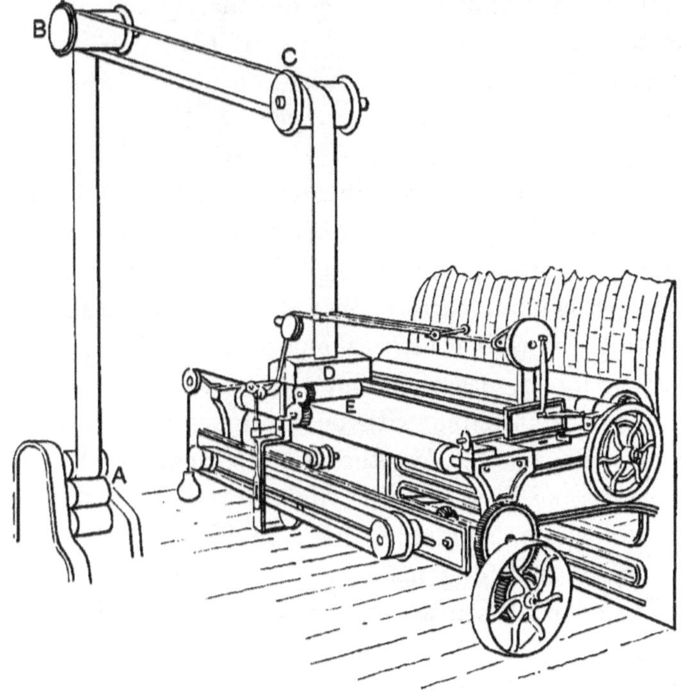

Fig. 21.—Scotch Feed.

the duty of the old bottom card in the hand-carding against every worker card one by one as it passes, so that in one revolution it has, on account of its greater flatness of surface, more carding duty to perform than all the workers combined. We call one of each of the pairs of top rollers a "worker" card, in distinction from the adjoining one, which is a stripper, but where is the duplicate of the old bottom hand card ? It is a mistake to infer, as some do, that the main cylinder is not much of a "worker" card, and that it only conveys the wool from one part of the machine to another. In carding, two worker cards are always required, working with points face to face; therefore, the main cylinder always represents the old bottom hand card, and consequently has something to do besides merely conveying the wool from one top worker card to another; it is itself the bottom worker card against every top worker card around the circle, as well as the conveyer of the wool and the stripper of all the strippers.

How many machines constitute a "set" has to be determined entirely by the class of work that the "set" is intended to be employed upon. If a good quality of all-wool work is intended to be manufactured, a full-sized set, consisting of scribbler or breaker, containing two parts and breast, scribbler, clearer, or intermediate of two parts, and carder or last machine, with one cylinder and six workers over, should be used. For other descriptions of work a "set" may consist of first machine of two cylinders and breast, an intermediate machine of one cylinder with five workers over, and a finishing machine of one cylinder with six workers over. Then again, a "set" may consist of first machine containing two cylinders and breast, and no intermediate but only a finishing machine of one cylinder with six workers over. This last kind of "set" is mostly used for wefts composed of mungo, waste, and a portion of wool. The usage is very different in different districts.

In the west of England the "set" of carding machines consists of three single machines of one cylinder each, without any breast-part. There is no question about our ordinary sets of carding machines being much too lumbering. The single machine system is much neater, and may be

made quite as effective if the large cylinder is raised
sufficiently upon high pedestals fixed upon low frames, so
as to give surface-room enough to place at least six workers
over each cylinder; such a set, clothed with suitable cards,
takes less room, less power, makes less waste, and is in every
way better, and costs less. But in cases where people are
still partial to the old breaker scribbler, with its two parts
and breast, the intermediate, as well as the last machine,

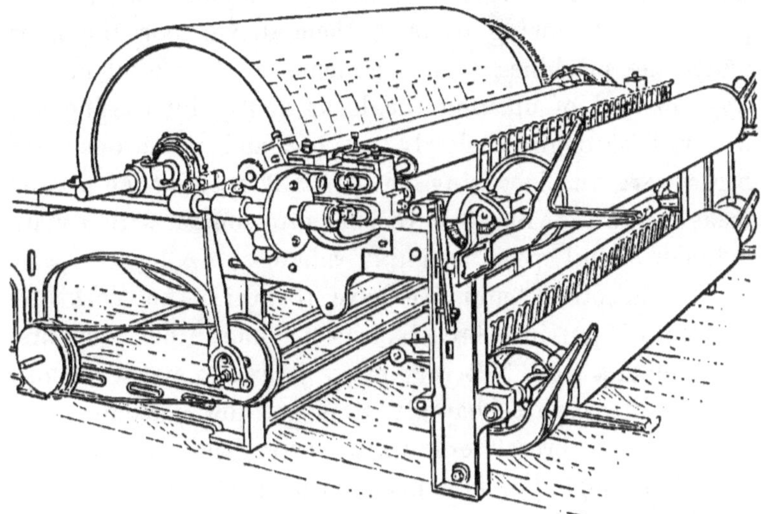

FIG. 22.—SINGLE STRIPPER CONDENSER.

might with advantage be single cylinder machines covered
with six workers. For weft purposes, in many instances,
a first or breaker scribbler, of two parts and breast, with a
finishing machine of six workers, without an intermediate
machine, makes up a nice set where only light working is
required.

In Yorkshire, until recent years, every carding machine
had a breast-part attached to it; but in the west of England no breast-part was attached to any of the machines,

not even to the first machine of a set. This latter custom is even more absurd and irrational than the Yorkshire custom, on account of the damaging effect that follows the feeding of the wool in the lock direct to the main cylinder revolving at full speed. The effect of this savage treatment of the wool is simply that a large percentage of it is at once ground and converted into mungo, in consequence of the full-speed cylinder grinding some of the more matted of the locks and snapping others into pieces and throwing parts of them down upon the floor under the machines. To commence carding wool in this way, instead of opening the locks gently by passing the material through a slowly moving and more open set breast-part, and loosening and partially opening the matted locks before allowing them to come into contact with a main cylinder at full speed, is simply ruinous.

On the other hand, we have to notice the Yorkshire custom of putting breast-parts to more than the first machine of a set. Not very many years ago every machine had its breast-part; happily, it is not now so common, as there is no breastwork to do after the material has passed through the first machine of a set. But whatever number of machines are decided upon as, constituting a set for any kind of work, whether with or without breast-parts, one thing ought always to be borne in mind, and that is that the last machine of a "set" should be a single machine, and more especially is this the case when the machine has to be followed by a condenser. It is very undesirable that the carded wool should be subjected to the action of two fancies while passing through the last machine on its way to the condenser, as the tendency of every fancy is to throw the carded material, more or less, from the sides towards the middle of the machine, thereby

reducing in thickness three or four of the outside threads
on the condenser, on each side of the machine, causing
the middle threads to be proportionately thicker.   When
the last machine of a set consists of two parts, first
and second cylinders, this tendency to throw off fancy
from the sides to the middle is doubled, and the mischief
of unevenness, between the side threads and those from
the middle of the machine, is aggravated twofold.   Much
neater and much better work in every way can be
obtained from a finishing machine with one cylinder
covered with six workers than from a machine consisting
of two parts or cylinders covered with four workers
each, in which the work has to run the gauntlet of two
fancies.   In addition to avoiding the mischief of two
fancies there is the further advantage in having six or
seven workers over the last cylinder for finishing, as with
such a number of workers over the finishing part, the wool
can be better and more uniformly distributed on the
surface of the last cylinder, and better prepared for doffing
evenly, especially if four of the first workers are made to
traverse in couples.   Say that the first worker starts out
towards the right hand for the space of three-quarters of
an inch, then the second worker must have started out
towards the left hand precisely at the same time, and must
complete its traverse out and in again, ready to start on
another journey at the same time as its companion, the
first worker.   The traverse out and in again must be
performed twice over while each worker is revolving once.
The two workers are best coupled together by a light bar
swivelled from the middle, with the stud attached to the
side of the bend of the machine, and then one traverse
pulley will work them both, and it will furthermore
ensure their always travelling in opposite directions,

which is a primary condition of success. Let the second two workers be arranged precisely the same as the first two. It will be seen from this that the flange of the traverse pulley will require to be in four segments, and must deviate three-quarters of an inch from the centre to the extreme of one side, and then to the contrary extreme of the other side. Of course, the collars of the workers that traverse will require to be turned off, so as to allow of the workers traversing. The workers move all the steadier if a lever about 2 feet long is fixed to the frame by a stud at one end and a weight at the other, while a double-flanged chain pulley rests on the worker chain, and is fixed by a stud to the middle of the lever. This simple little lever acting on the worker chain can be placed inside the frame, and as near as possible to the floor; it does away with all that "jerky," fitful motion of the workers, and all move steadily as the lever rises and falls, just as the workers sway backward and forward, and all slack chain is taken up and given out again as required. The last two workers do not require to traverse, but move in the ordinary way. There is a great advantage in having the first four workers traverse, whether the doffer is a ring doffer or a fillet doffer, for the Bolette or Martin condenser with its dividing tapes.

After the carded wool leaves the last doffer of the set of carding engines in our modern mills it is received by the condenser, of which there are a variety, and for each of them some special claim to attention is put forward by their respective advocates. Some claim for their machines that they can turn off more work than others, and that therefore they have prior claim to notice on that account; but all such claims are utterly beside the mark, as the condenser is only an intermediate machine amongst a series of machines, in a series of processes, and has nothing whatever to do with the quantity of work turned off. The *condensing is only a mode of doffing* and passing on of the carded wool to the spinning, and the condenser can only pass on the amount of carded wool that a set of carding machines are able properly to card, and no more, so that the quantity of work turned off does not depend upon the condenser, but upon the carding capability of the set of machines in use, for the time being. Any condenser in general use will be able to pass on to the spinning as much carded wool as is presented to it, and whatever the merits of any particular condenser may be, it can do no more than this, it can determine neither the quantity nor the quality of the carding process of the work in hand. The kinds of condensers may be classed under four general heads, viz. :—

1. Single doffer and double stripper machines.
2. Single doffer and single stripper machines.
3. Double doffer and double stripper machines.
4. Fillet doffer lap machines with grooved metal dividing cylinders and tapes, such as the " Martin " and " Bolette " machines.

The first-named style of machine has been most commonly used, but the second is now mostly in favour; the third, the one with two doffers, is so extremely vicious in principle that no one with any insight into its mode of action will adopt it. The upper doffer, being the first taker of wool from the cylinder, has the first choice, and sorts the wool to such a degree that the bobbins from the upper doffer will very rarely spin with those from the lower doffer, but have to be spun by themselves.

The single doffer principle is the only one that can be adopted with safety, and the single stripper is the best in principle also; but with some kinds of material the rubbers are sometimes a little difficult to manage, as their length of stroke is so circumscribed. The one doffer and one stripper is the simplest kind of machine, whether for ring doffers or for fillet doffers of the Bolette and Martin's patent kind. We are indebted to Messrs. John Haigh and Sons of Huddersfield for tracings of end sections and descriptions

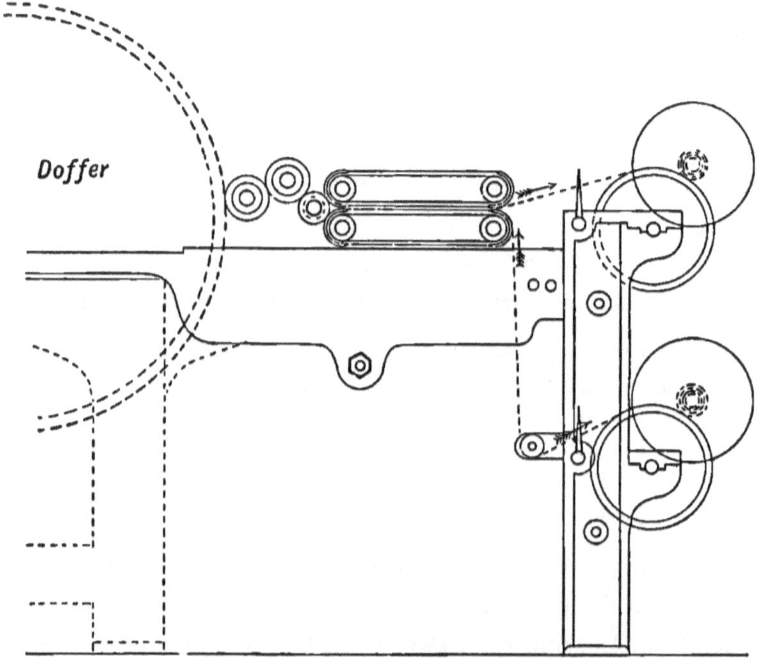

FIG. 23.—END SECTION, SINGLE STRIPPER CONDENSER.

of three different kinds of single stripper condensers, Figs. 22, 23, 24, and 25. The mode of dividing the lap from the fillet doffer of the Bolette and Martin's patent Belgian machines is shown at Fig. 26, which is by large metal grooved rollers, the ridges of one roller working into the grooves of the other, as seen in the figure. In the ordinary ring doffer machine one thread is taken off for every inch

of width on the wire, after allowing for a waste thread on
each side of the machine, so that each ring is thirteen-six-
teenths wide, and the other three-sixteenths is the space.
This width of ring during recent years has been very unwisely
reduced, of which we shall have more to say presently.

The writer has produced yarns for Cheviot fancy
woollens from which he has drawn fibres of wool 6

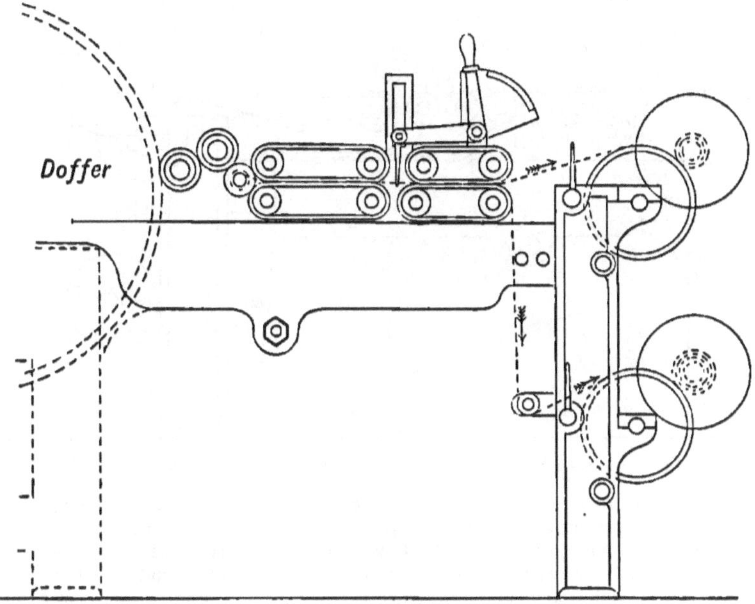

Fig. 24.—End Section, on the Tandem Principle.

inches in length from the condenser sliver after the
material had passed through a set of carding machines and
the condenser, so that in working such material the above
width of rings and spaces were found quite narrow enough,
to say the least. A ring nearly double the width and
quarter-inch spaces would have been much better. It is
very obvious to a thoughtful observer that a long tethery
wool would be extremely difficult to divide from the lap,

either by the Bolette or Martin machine, or any other of
its class, whether worked by one entire endless strap or by
a series of endless straps, one for each ribbon.   Clearly
the only mode of dealing with such long wool is not to
allow it to become interlaced into a web or lap, but to strip
or doff it directly from the carding machine in ribbons, by
means of a ring doffer.   Obviously such a wool as we have

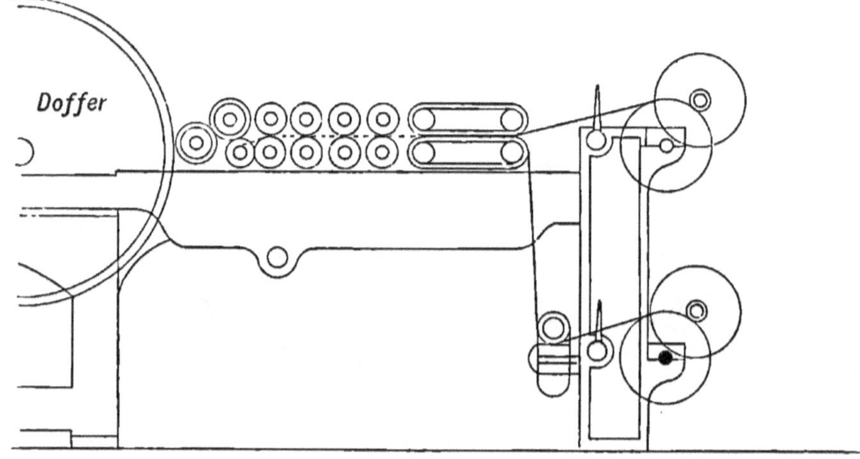

FIG. 25.—END SECTION, BOOTH AND BEAUMONT'S PATENT PRINCIPLE.

This condenser consists of two stripping rollers, one divider roller, four pairs of
rubber and drawing rollers, and one pair of rubber leathers, 6 to 8 inches centres.
The advantages are, a longer staple of material can be worked than on the ordinary
condenser, also by having a draft between each pair of rollers, and a further draft
between the rollers and the rubber leathers, from 20 to 25 per cent more length can
be got in sliver.

just alluded to would be unmanageable if once allowed to
form into a web or lap, and yet such wools are frequently
used for clothing purposes in the fancy woollen districts of
the United Kingdom.   No condensers other than ring-
doffered ones ought to be set to work upon material of this
kind.

If manufacturers will still persist in using the condenser,
the best machine, in the writer's opinion, is the one doffer and

one stripper ring machine; and as there is some difficulty in rubbing some kinds of material on the one stripper machine, on account of the shortness of the stroke, this might be surmounted by having an additional pair of rubbers to the top bobbin, and a like additional pair of rubbers to the bottom bobbin; this method would allow the two pairs of additional rubbers to have a full stroke, as each pair would only take every alternate sliver as it passed on its way to its respective bobbin.　By using a condenser with only one doffer covered with rings and one stripper we

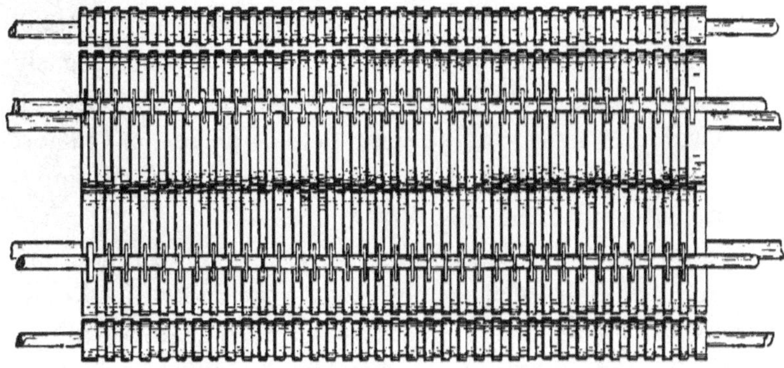

FIG. 26.—BOLETTE DIVIDING CYLINDERS.

ensure the greatest evenness or uniformity between thread and thread in point of thickness, and the first pair of rubbers could be devoted to the completion of the dividing process, and leave the bulk of the rubbing to be done by the two additional pairs of rubbers.　In the one stripper machine the stroke of the rubbers has to be very short, to prevent the slivers from rubbing into one another; but even this short stroke could be shortened, and the pressure of the top rubber reduced, and the rubber made narrower, if these first rubbers were only used to assist the dividing roller and to partially rub the slivers, so as to give them consistency to carry themselves forward to the additional pairs of rubbers to have the rubbing completed.

It is worth while to incur a little extra trouble in the rubbing in order to obtain evenness in the work, and this can only be attained by doing the doffing all at one place or in one line by a single ring doffer, and by doing all the stripping at one place and in one line right across the machine by a single stripper. We think that these are the only lines on which we may expect to obtain even and uniform work. Some of the above thoughts and remarks occur in a notice of Bolette's condenser, which appeared in the *Textile Manufacturer* for April 1880. Machines of the Bolette class cannot be advantageously employed upon such long wools as those mentioned on page 167, but must confine their field of operation to short wools, and in the main to fine wools, and leave the long-fibred material to the ring doffer.

There are many kinds of feeds at present in use, but judging by their relative popularity, the Scotch feed (Fig. 21) seems during recent years to have come most prominently to the front. It is very simple, and, taken all in all, is a very useful and fairly effective feed, and, with the additions and improvements lately made in it, it has become practically the most effective working feed. The side drawing-off motion gives the material a crossing of something like sixtyfold. The patent improvement by W. Lawton has greatly added to the efficiency of the Scotch feed by gathering up the edges of the sliver, making them much more compact and solid, and therefore much stronger than formerly, and consequently much less liable to break down in passing from one machine to another. In Lawton's patent apparatus there are two little shufflers, one on each side of the sliver, just before it

leaves the travelling lattice, which thrust in the ragged
edges, and make the sliver compact just as it enters the
drawing-off rollers.

A is a small revolving shaft or spindle, and is driven
from the fancy end in the scribbler with a band or belt,
and runs in two bearings. On this shaft are fixed two
eccentrics, B, which are set at opposite centres. C is a flat
groove or slide, in which are placed two dies, D. These

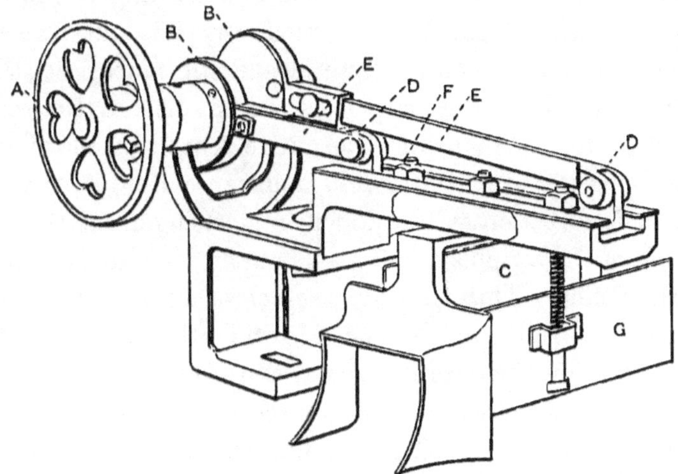

FIG. 27.—LAWTON'S PATENT.

dies are connected by two rods, E, to the eccentrics B,
which causes them to be moved backwards and forwards in
the slide C in such a manner that they are both in and out
together. Through the dies D are fixed two set-screws, F,
to which are suspended the two guide plates G; these are
kept in continuous oscillating motion, and between them
the web or sliver passes. This little apparatus can be
fixed to any existing machine in two or three hours. The
flat sliver as it emerges from the drawing-off rollers is
carried up and overhead, and then let down into a runner,

that lays it on the feed lattice of the next machine in folds from side to side, which overlap each other about half the width of the sliver at each succeeding fold.

The next in popular favour is the Apperly harp feed, which traverses the sliver across the feed lattice much after the manner of the Scotch feed, but lays the sliver diagonally on the feed lattice, as its name implies ; and instead of the sliver being flat and ribbon-like, it is round like a rope. In using the Apperly feed in the west of England, where it originated, a great deal too much twine is put into the rope, much more than is required, and still much less twine will suffice if carding engineers would place a tin roller about 7 inches diameter the entire length of the doffer, just opposite the comb-stock, and cause it to revolve in the same direction as the doffer by a string from the end of one of the strippers, and at about the same surface-speed as the doffer. This tin roller, set close against the sliver, will help to roll it on the one side while the doffer rolls the other, and by this means a round and neatly formed roping can be obtained without that excess of twine that is at present put into it. By this little device a firm roping can be obtained that will bear carrying overhead in any direction without breakage, and can be fed into the next succeeding machine without breaking up into patches through being girded into a hard rope by having in it an excess of twine.

What causes people to resort to these two feeds so freely is that they are so soon run off at the finish of a parcel, and run on again at the commencement of a new one. This is a great convenience where small parcels are common, as is often the case in connection with the fancy woollen manufacture. Some of the bank, lap or blanket, and other feeds are often so long in running off and on again that part of a set of machines has to stand idle for the greater part of a day, waiting for the others to finish ; and on the other hand, in commencing a new parcel of work, the first machine requires to run some considerable time before the others can be got work. In the case of the two feeds mentioned, they can be run off and on again as quickly almost as when the machines were fed by hand, which is a point of considerable importance when "finishings" are frequent. The Marsden and Blamire's lap feed is a most excellent one where there is room enough to place it, and where it can be kept at work upon large parcels of material, and is not required to "finish" every alternate day. The Scotch and Apperly feeds take up so little room, that this often settles the question of choice, though where there is a sufficiency of room, and where "finishings" are not so frequent, the Marsden and Blamire's feed is perhaps the most perfect feed yet invented, as its crossing and mixing capacity, when skilfully managed, produces the most even kind of work hitherto obtainable. The choice of feeds will in many cases very much depend upon the spare room and the kind of work that the feed is likely to be chiefly engaged upon, but where people are well informed as to the merits of the different kinds of feeds and are not pressed for space, they should use their discretion and not leave the kind of feed to the recommendation of the machine maker employed to make the set of machines. The carding engineer, to have an intelligent opinion of his own, ought to make a study of the different feeds for himself, and not be at the mercy of others.

Another matter that demands some notice in passing is the material of which the machines should be built. Of late years the card-bearing surfaces have in many instances been of cast-iron. There are many grave objections to the

use of cast-iron for this purpose. The thing to be aimed
at is to build the card-bearing portions of our machines in
such a manner and of such material as will not warp
or swerve from truthfulness of outline when subjected
to varying degrees of temperature or moisture by the
change of the seasons of the year; and the second
consideration is that the material used shall have no injurious effect upon the cards. After all the different
experiments that have been made in this line we strongly
incline to having all the cylinders, and nearly all the
rollers, made of wood, but not thrown together, as it were,
in that rough and unscientific manner of long logs reaching
the whole transverse of the machine, but to use the wood
in the most scientific manner known at the present day.
We can obtain all the durability and all the truthfulness
of outline which is so essential an item in a carding
machine by building of wood, if we only set about using
the wood in a scientific manner. Instead of building
our cylinders of long transverse logs, as woollen machines
used to be built, let them be built of small segments,
say about six inches in length, and about three-quarters
of an inch or an inch in thickness. By this means we
can obtain all the lightness and durability we require,
and last, but not least, a foundation upon which to fasten
down our card-cloth, much more convenient and much less
liable to corrode, stiffen, and destroy the backs of our
cards than cast-iron, and of much less weight and consequent waste of power in driving our machine. There
is, however, one kind of rollers in the machine that we
would recommend to be made of iron, and that is the
stripper rollers. They are the smallest of the roller tribe
in the machine, and when well and truthfully made
of iron act as "straight edges," and point out immedi-

ately any variation from truthfulness, should any occur in the working portions of the carding machine. There is another point of advantage to be noted in having the card-bearing portions of the machine nearly all of wood, and that is that wood tends most to the durability of the cards. So delicate a thing as a scribbling card placed upon an iron surface foundation soon corrodes, and stiffens so as to become useless for all good working purposes, whereas by placing our cards upon smooth, well-seasoned hard wood surfaces we ensure their safety from corrosion for years ; indeed, the difference in favour of wood surfaces, in the durability of the cards alone, is so great that it cannot be lightly passed over in the economical management of a large concern.

The kind of cards and the degree of fineness are matters that must be determined entirely by the speciality of the manufacture they are intended to be employed upon. There is no advantage in having cards too coarse for their work. If we err at all in this matter, we should be careful to err on the side of fineness. All that can be done here is to give a general specification, because each carding engineer will order cards according to the kind of work for which they are intended. The following specification for cards would be available for a good quality of all-wool goods, or for wool and an admixture of mungo. If the material should be of long staple, and coarse, too fine cards would soon be worn out, and would have to be replaced by others of a stronger kind, more adapted to long coarse material. The following may be taken as a general representation of a set of cards for a good class of work. Blends of wool and mungo cannot be worked to advantage upon coarse strong cards.

CARDS FOR A BREAKER SCRIBBLER, TWO PARTS AND BREAST

The feed-rollers of strong needle point, leather filleting, licker-in, short cut.

	Counts.	Crown.	Size of Wire.
Angle-stripper	60	6	25
Breast Cylinder	70	7	25
Workers	75	7	25
Strippers	70	7	25
Fancy	60	5	25
Doffer	75	7	25
Angle-stripper	65	6	25
First Cylinder or Swift	100	9	30
Four Workers	105	9	30
Four Strippers	80	8	30
Doffer	105	9	31
Fancy	70	6	31
Angle-stripper	75	7	31
Last Cylinder or Swift	130	11	35
Four Workers	135	11	35
Four Strippers	100	9	34
Fancy	80	8	36
Doffer	130	11	35
Cards for intermediate Scribbler—one Cylinder—			
Angle-stripper	80	8	25
Cylinder or Swift	130	11	35
Five Workers	135	11	35
Five Strippers	100	10	34
Fancy	80	8	36
Doffer	135	11	35
Cards for last Machine or Carder—one Cylinder—			
Angle-stripper	100	9	32
Cylinder or Swift	135	11	36
Six Workers	140	12	36
Six Strippers	110	10	35
Fancy	80	8	36

Another matter at this stage is the grinding of the new cards when first nailed or put upon the machines. In modern mills it is usual to have a grinding frame the width of the widest machine in the mill, and then, by the aid of a loose sliding head-stock, any narrower width of machine can be accommodated. The best grinding is that which produces a point nearest in form to that of a needle point, and when a good ground point is once obtained it can be retained by the action of card against card in the working, as a card is the best sharpener of a card, when the machine is once got fairly to work. The stripper, properly set, keeps the worker in full point, and the fancy, properly set to work sufficiently deep into the large cylinder or swift card, will keep the point in good working order. The angle-stripper will keep the doffer in point, except the last doffer, to which there is no angle, and which frequently needs a card-roller to keep it in point. Some carding engineers use a small emery roller, but a card-roller gives a better point. Grinding should be very sparingly indulged in after a newly-clothed machine has once got the new cards fairly to work. The Continental plan of a weekly grind is false in principle and mischievous in practice; it is utterly unnecessary if the cards are properly set; it is needlessly grinding the wire to waste that ought to be reserved for carding the wool, and using up a set of card clothing in three years' time, or less, that ought to last six years, to say nothing of irregular work in consequence of the point varying from day to day. A very great deal depends upon proper carding in the woollen manufacture, and anything that tends towards greater uniformity is of the utmost importance, and every precaution is necessary in order that the sliver may leave the carding machine with the greatest regularity from hour to hour and from day to day. There is no such thing as combining two or more slivers together in woollen, and so reducing a fault in any individual sliver by distributing it amongst others. Every individual sliver has to stand on its own merits, and if it leaves the carding machine with a fault, that fault sticks to it, and produces a faulty thread, which remains faulty through every succeeding process to the end of the manufacture. We know of no other textile industry that is so peculiarly circumstanced. This being the case, the intelligent manufacturer feels himself constrained to use every effort to have the carding department skilfully conducted. An unskilled workman in this department is very dear at a gift, and entails more loss by the misuse of the cards, many times over, than any difference in wages amounts to between unskilled and the most expert and best skilled workman that can be obtained. Proper carding, then, includes properly selected cards for the work they are intended to do, seeing that they are properly placed on a good dry wood foundation, so that they will not be liable to corrode before they are half worn, proper grinding and proper setting when all is ready for commencing work. There are a variety of emery rollers for grinding frames. Some of the rollers are corrugated on the surface in order to catch the sides of the points of the card teeth in grinding. Any kind of emery roller ought to traverse from right to left, and the contrary, so as to counteract the tendency to grind the wire to what is termed a "chisel" or a flat point. Many of the emery rollers scarcely traverse quickly enough to accomplish this object. In grinding it is never wise to force the work forward too quickly, because if the grinding is forced forward too quickly, there is always a liability to grind the point into a hook at the extremity, in consequence of the excess of pressure, and a hooky point is always very troublesome when the cards are set to work, and should always be avoided by allowing sufficient time for the grinding operation, and by not putting the emery roller too keenly on the card. The best ground point that can be obtained is only a make-shift affair, so as to get the cards to work until they can be brought to act upon each other during the ordinary working of the machines, which action produces a needle point far superior to any ground point for all working purposes. When all has been done that can be done by machine grinding, it is frequently necessary to get a good sheet hand strickle, and pass it very quickly but lightly over the surface of the revolving fancy from right to left and *vice versâ*, and sometimes other rollers also, while still in the grinding frame, to take off the roughness, jagged edges, and hooky points of the cards, before finally putting them into their places for work. The last operation requires to be deftly and rapidly performed by a skilful hand of light touch.

During late years steel wire has come into much more general use for card-making than formerly, and deservedly so, as it is capable of a better point, and keeps the point better; it wants handling with a little more care, and though it costs more to commence with, it lasts longer, and with judicious management does its work better. The only exception to the use of steel wire would be in the rings for the condenser doffer; a more uniform ring can be made in the old iron wire than has hitherto been done in steel, but this difficulty will, no doubt, be overcome before long.

There is something startling about the setting of cards, and something that may be regarded as singular in this nineteenth century of mechanical enlightenment. We mean here by "setting" the placing of the respective rollers and cylinders at their proper distances in respect to each other, so as to card or open the wool properly. The Red Indian, whom we call a savage, uses all the senses that he is possessed of—sight, hearing, touch, etc., in the pursuit and hunting down of his prey; but the civilised man—the carding engineer—will frequently only condescend to use one of his senses, the sense of sight, in setting his cards. In England, till within a very few years, the carding engineer only deigned to use sight in setting his cards, and this is all the more marvellous as the distance to be judged frequently amounts to very little more than the one-hundredth part of an inch. The Continental carding engineer uses a gauge, which is a thin steel plate about an inch in width, sometimes a little broader, and 10 or 12 inches long, which he gently slides in between the cards, and measures the distance by an actual measuring instrument as well as by sight. It is only quite recently that the English carding engineer has condescended very reluctantly to use a setting gauge in adjusting his cards to their working positions.

In the matter of card-setting almost every carding engineer you come across has his own particular "fad," so little headway has thorough technical knowledge yet made in this matter. A higher technical knowledge is the urgent want of the age in many departments of our woollen manufactures. The setting of cards by sight only precludes our getting the full amount of work out of the machines that they are capable of turning off. By the use of the setting gauge exactness can be attained, as well as the utmost safety, and we get also the greatest amount of work out of the machine of which it is capable. By the old mode of setting by sight, we get little more than half the amount of work out of our machines that they are capable of yielding, and as a consequence we have to order machines almost twice as large as is absolutely necessary, and thereby involve more first cost, and take up more room and power than is otherwise necessary. Take a steel setting gauge and go round a machine that is just newly set in the best manner possible by sight only, and you will probably find that the first worker you test by the gauge will run safely almost a third nearer the cylinder or swift; the next you will in all probability find is too near, and could not be run without losing point; the next stands badly to the light, and is twice as far off the cylinder as is necessary; the remaining workers are passable; the strippers are found to be quite as irregular as the workers. Try the doffer; not near enough to the cylinder. Go round to the other side of the machine and try the doffer on that side; twice as far off as it ought to be, and the workers down the same side of the machine are many of them quite the opposite in nearness to the same workers on the other side of the machine. "But why all this difference?" asks some one. But how can it be otherwise? The light falls differently upon different parts of the machine, and the sight alone cannot measure those minute spaces with sufficient exactness in an ever-varying light. The best sight-set machine will show these irregularities the moment it is tried by a steel gauge. Why limit the setting to sight and hearing when touch can be applied through the steel gauge with so much more certainty? The steel setting gauge is an absolute measure by which every working part of the carding machine can be adjusted exactly to the position desired. Why, then, should our carding engineers block the way of the progress of our manufactures by refusing or neglecting to use the requisite tools for their daily calling? forgetting or disregarding the teaching of one of our commonest everyday proverbs, that "Neither workmen nor fools can work without tools." The carding engineer cannot possibly work his machines properly without a setting gauge. The thickness is stated in the numbers of the wire gauge, and a different thickness of gauge has to be used for setting different parts of the carding machine—a No. 24 gauge for the forepart of the breaker scribbler, or for some kinds of work a No. 20, then gradually increasing in closeness of setting until on reaching the last machine a No. 30 to 33 may have to be used. No. 31 on the Birmingham wire gauge is about No. 33 on the gauge in general use in Yorkshire.

In respect to speed, 650 feet per minute is fast enough to run the surface of a breast cylinder, and 1000 feet per minute for a first large cylinder or swift, and say 1050 for the other large cylinders to the end of the set; but speed must depend entirely upon the class of work in hand—that which would be right and proper for one kind of work would be very unsuitable for another kind; the above rates would only be suitable for good all-wool work of good quality, or where there was only a small admixture of mungo along with wool. The writer has used camel-hair noils and alpaca noils along with wool in making yarns for certain classes of fancy woollens, where the speed has had to be reduced to half the rates given above, so that the student will understand that speed in woollen-carding is a relative term, and has always to be considered in its relation to the kind of material, and the skilful carding engineer will always adapt his speed to the material in hand. It is absurd to think of attempting to run machines on loose coarse material at the same speed as if the work was of a very fine Botany quality. Then, again, there are kinds of work where the middle doffers and the workers require speeding up, so as to get the work more quickly through the machines, to cause the material to work more thinly on the cards, and thereby make less waste when the material is of such a nature as not to require much work, but an intimate mixing together with very light working on the cards. No such thing as a hard-and-fast line can be laid down; a skilful workman will always be careful not to overrun his machines, but will always have due regard to the nature of the work, and will regulate the speed accordingly; a specification of speeds that would be very appropriate for one kind of work would be very widely astray for another. No statement of speeds, either for a whole machine or for any of its individual parts, can be given that will dispense with the constant need of judicious supervision and clear-headed judgment on the part of the skilled carding engineer, who will require in each instance to take into account the condition of the cards and to use his best judgment as to whether any increase or decrease of speed in the whole machine, or in any part of any particular machine, will improve the work on which for the time being it is employed. Speed must always be tempered to the work; it is one of the conditions of good workmanship. Another condition of good workmanship is that the respective parts of the carding machine be duly proportioned to each other. The breast cylinder should not be less than 40 inches in diameter, the large cylinders or swifts 50 to 52 inches, the workers not less than 9 inches, the small strippers $4\frac{1}{2}$ inches, and the doffers 32 inches. By having the doffers and workers of good size, the carding engineer is much better able to preserve the staple of the material upon which he is at work; by having large-sized doffers the workman can run them at a much quicker speed and still clear the cylinder more thoroughly of its carded wool, as there are more card points presented to the cylinder at the same time than is possible in a smaller diameter of doffer, and the wool works much thinner on the machine, and by being less clogged and crowded the staple gets less broken, and much less waste is made. The same remark applies to larger workers; there are more points presented to the cylinder, and as a consequence fewer fibres for each point, and the staple of the material is much better preserved than it can be on a small worker. A small diameter in any of the respective working cards is a thing to be avoided, as it means broken staple, more waste in the carding, and a crumpled-up, rough-looking weak yarn. Any old machines with small doffers, if in otherwise good condition, as the cards wear out may be much improved by clump-covering the doffers, and so increase the diameter 3 or 4 inches before re-covering with new cards.

We must not close our remarks on carding without to some extent alluding to the making of the cards, and the form in which they are made. It is a very important matter, and yet in many respects it receives very little thought in comparison to its due, and really seems to be very little understood. A properly made card will turn off 20 to 30 per cent more work, besides doing its work much better. The form of the tooth, and the angle at which it is bent, have very much to do with the card's good working capabilities. After the tooth has been bent back in the lower portion or shank, in bending it forward again to make the working hook, it should not be bent any further than just to touch a perpendicular line raised from the point or place at which the tooth enters the foundation leather into which it is set, as is shown in Figs. 38 to 47, where a variety of forms of card teeth are to be seen, that are all brought into such a position as to be faceable to a perpendicular line. The object of this perpendicular

line illustration is to show that whatever form, or style, or
degree of bend we choose to adopt, it must always finish

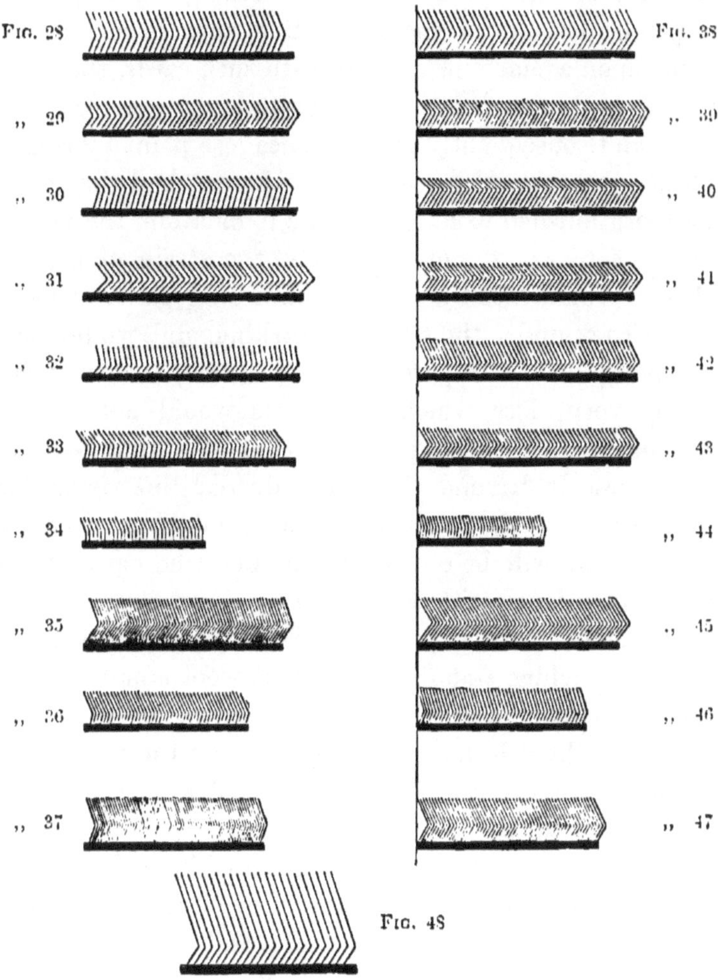

Fig. 28

,, 29

,, 30

,, 31

,, 32

,, 33

,, 34

,, 35

,, 36

,, 37

Fig. 38

,. 39

,, 40

,, 41

,, 42

,, 43

,, 44

,, 45

,, 46

,, 47

Fig. 48

VARIETIES OF CARD TEETH.

at a point exactly over, or perpendicular to, the point at
which it started, or where the tooth rises out of the foundation. If this condition is not always kept in view and

attended to, the card cannot work properly. The reason why such a rule as this has to be laid down is that when the bend of the tooth we choose to adopt goes past the perpendicular line of the foundation, then as soon as wool is put upon a machine clothed with such cards, the points rise as soon as ever they feel the strain of the wool upon them, and consequently must at once lose point by running into the next neighbouring card; whereas, had the bend of the tooth finished exactly over the foundation, the point of such a card would not rise under the strain of the wool when it commenced working, but would rather sink by falling back under the strain of working, and would never lose point when properly set, and would, moreover, bear setting very close, which the others would not. On no account ought a working card to be made with a bend that goes forwards beyond the perpendicular line rising from the place where the tooth enters the foundation. The need for this rule will be obvious to any one who has any practical acquaintance with the working of carding machinery. Never have a card that will be liable to rise after setting with the machine standing as soon as wool is put upon the machine, as such a card will never bear setting close up to its work without losing point and injuring the points of all the other cards that are working against it, especially if it is a cylinder card.

The form of card shown at Fig. 28 makes a fair average good working card, and will bear very close setting up to, and get through plenty of work, and will do it well. If a card a little keener in the hook is thought desirable, then try the form of card shown at Fig. 31, which is a keen-set point with the bend in about the middle of the tooth. It will do its work exceedingly well if handled with skill and care, more especially if made from steel wire.

The form and structure of the woollen card has been about as little studied, or studied to as little purpose, as the structure of the woollen thread, as we shall learn further on, and the result is that we have been getting longer and in every way larger machines, which not only cost more at first hand, but cost more in room and power every year, and turn off inferior work. A set of two cylinders and breast for first machine, one cylinder for second or intermediate, and one cylinder for last or finishing machine, is amply sufficient for very high-class all-wool work of first-class quality, whether for plain or fancy goods, if only covered with suitable cards that will bear setting up to their work, and that are capable of working when they are set, instead of having such long lumbering machines, covered with cards, that will neither bear setting sufficiently close to their work, nor work to any purpose when they are set. A card with the upper portion of the tooth carried past the foundation line cannot bear close setting, neither can it get through any amount of work, nor do it properly, and hence we have continuously added to the size of our machines, instead of clothing the lesser machines with cards of proper form and with full working capabilities that could turn off both quantity and quality of work. The Figs. from 28 to 37 show a variety of forms of card teeth of various working capabilities, but whatever style or form of card we select, we should always be careful to instruct the card-maker to bring the point of the tooth no more forward than will be faceable with a perpendicular line raised from the foundation at the exact point or place where the tooth enters the leather, or where it rises out of the leather or other material into which the tooth is fixed, as is illustrated in the Figs. shown on perpendicular line from Nos. 38 to 47. Fig. 48 is an illustration of a steel wire fancy about 70s. 7-crown, 2-inch cut. We often do a great deal of mischief by having our fancies too full of wire; it should always be borne in mind that the fancy is not a working card, but a stripping card, and that of a very peculiar kind; so peculiar is its form and the mode of its action that it actually becomes a brush stripper, or a stripping brush, as it does not present the face of its point to the cylinder card, but only the back. The ordinary stripping card presents the face of its point to the back of the point of the worker card that it has to strip, but the fancy presents the back of its point to the back of the point of the cylinder, and operates only as a wire brush, and in order to do its work efficiently it requires to work into the cylinder card to some considerable extent, to get out the carded wool completely, and to keep the cylinder card up to one uniform degree of keenness of point. When the fancy is too full of wire it cannot work into the cylinder card, but simply floats upon the surface, like a grinding roller, and instead of emptying the carded wool out of the cylinder, it only treads it further in and chokes up the mouth of the cylinder card, spoils the work, and wastes the material. *A fancy too full of wire defeats the object for which it is used.* A 70s. 7-crown steel wire, long cut, like the one shown in Fig. 48, will do all that a fancy is required to do, and this number never needs to be exceeded in working the finest cylinder that is made. Always remember that the fancy is not a working card but only a brush stripper, and therefore ought to be open set. Another thought in passing, and that is that it is never a wise policy to keep fancies on too long; as soon as they become hard and boardy to the touch they ought to be taken off, as they will soon take more than their value out of the cylinder cards if kept on after they have become hard and stumpy; another inducement to take them off in time is that a steel fancy like Fig. 48 when worn down makes a capital angle-stripper, and can be used in this way in many cases up to the bend, so that taking them off in time is no loss.

Even in the ordinary working card there is such a thing possible as getting a card too full of wire so that there is not room enough for the wool to get properly worked. There is a growing disposition amongst many carding engineers to order cards, say 12-crown one way and 140s. the other way, instead of 12-crown 120s., thereby edging in a little more wire, as it were, by stealth; but what we now most need is not more wire but a careful study of the proper form of the teeth of the card and the best way of using it; therein lies the efficiency, and for want of proper attention at these two points we lose at least 20 to 30 per cent of the carding power of the wire we already have. When a cylinder is too crowded with wire it only makes it more difficult for the fancy to clear it of carded wool. The cylinder card cannot work up to its full carding capacity when the fancy is unable to clear it thoroughly of carded wool as rapidly as it comes forward.

Figs. 29 and 30 are forms of teeth after the manner of those used by the French. They have the bend two-thirds of the way up the tooth, which gives them a very keen hooking point, which is easily stripped by the fancy. This is very essential with the French, as they do not allow the fancy to work into the cylinder card, but only to do little more than fan the surface of the cylinder.

Fig. 31 is a form of tooth of a keen bend, and with the bend only about half-way down the tooth. This form of tooth would work well with skilful usage, and would yield up its carded wool freely to the action of the fancy; anyway, to say the least, a card with less keenness of hooking point than is shown in Fig. 28 should not be used. Fig. 33 is too straight up in the tooth, and has little or no working power in it worth mentioning. Fig. 34 is altogether false in form, as it bends past the working line and is too straight up generally. The best forms of good working card teeth are to be found between Fig. 28 on the one hand and Fig. 31 on the other. Fig. 33 is a bad form of tooth, and only becomes workable when the bend of the shank, or bottom part of the tooth, is thrown back, as in Fig. 43. In the English card the bend is usually from half to two-thirds of the way down the tooth, and consequently requires to be moderately keen in the bend, or the card has no hook or carding power; on the other hand, the French have the bend only one-third of the way down the tooth, and yet they get a much more hooky point than the English with the same amount of bend, as the tooth is turned upside down, as it were, compared with the form of tooth commonly in use on this side of the Channel. Compare Figs. 30 and 31, also 36. Fig. 30 is bent back very little in the shank, and yet the hook of the point is keen, as the bend is such a little way down the tooth.

In an English card, on account of the bend being so far down the tooth, the shank, or part below the bend, has to be thrown a considerable way back before a keen working point can be obtained, and in practice it is not commonly done, and we err in not having a working point sufficiently keen, and resembling the point of the French card, to card wool in sufficient quantity as well as to be able to turn out the work well and efficiently done. The general run of our English cards are sadly deficient in working power. They have not sufficient grip and mastery over their work, and the wool trips off the points of the card and does not get sufficiently carded. A little more bend in the tooth, and the shank thrown a little farther back so as to bring the point into proper perpendicular working line at the finish, would be an immense improvement to the general run of our cards, and increase their working capacity at least 50 per cent, and the work would be much more satisfactory to all concerned, as the fancy would be able to clear out the cylinder cards much more easily and effectually.

In drawing our remarks on woollen-carding to a close, we would urge the importance of the operation upon the careful attention of the student, as very much of the welfare of the after operations of the manufacture is greatly dependent upon the carding operation being well and skilfully carried out. The machinery employed in the process is of a very expensive kind; the card clothing of the machines is also very expensive, and of a very delicate character, and soon injured, and therefore requires the utmost care and thought, in addition to great technical skill, in their management. Further, again, wool is amongst the most costly of our raw textile materials. All these considerations, both jointly and severally, call for the most careful attention, thought, skill, judicious watchfulness, and the fullest technical and scientific knowledge being brought to bear, and woollen-carding machinery ought to be placed under the management and care of persons who have been specially and carefully trained for the purpose, either by apprenticeship or other technical and scientific course of preparation. The education of both the head and the hand requires here to be carried to a high degree of excellence in order to discharge aright the duties of a woollen-carding engineer. If we think for a moment that as so much care and expense are incurred in the production of the machines and their delicate card clothing, a like amount of care and technical skill, combined with trained delicacy of handling, ought to be called into service follows as a rational and necessary consequence, as too much judicious care and trained skill cannot be bestowed upon an operation where so much is dependent on the issues.

PART X

SPINNING : ITS HISTORY, PRINCIPLES, AND PROGRESS

SPINNING, there is little doubt, originated in the far-back prehistoric times, of which we have little or no authentic information ; but the wants and necessities of mankind would early in the existence of the race force attention to the structure of textile tissues. The ordinary wants of everyday life would call into being threads or tissues of different kinds, however rude in structure, both vegetable and animal. Many things, no doubt, would be laid hold of, and twined and manipulated into a thread-like form to meet the requirements of everyday life. The first results would doubtless ill compare with what the present day turns out ; still men's observing and inventive faculties would make advances, and little by little the raw products would be manipulated into tissues or threads more or less resembling the textile tissues of modern times. The early nomadic tribes, as they wandered from place to place seeking fresh pasturage for their cattle and sheep, and making garments of the skins for their own covering, would invent and use threads of some kind with which to sew or fasten together the skins so as to make them into some kind of convenient wearable form. The wool torn from the sheep by bush and bramble would attract their attention, and easily suggest and lend itself to twining and forming into something like a thread. Be this as it may, the origin of the spinning of textile threads is historically little known, but we are warranted in believing that from the earliest times mankind have clothed themselves in the wool of the sheep, either in its manufactured state, or in its natural state on the skin. In his work on *Cotton Spinning*, Mr. Richard Marsden says that he "in another place has ventured to suggest that the first spinner was a shepherd-boy, and the material used a few locks of wool. Reclining under the shade of a tree whilst his flock was feeding around him, it might easily happen to a playful youth to have his attention attracted by a small portion of a cast fleece lying near, to which he would stretch forth his hand. Toying and amusing himself with this to relieve the tedium of the hours, it might quite as easily happen that he should twist its fibres together between his fingers, and, surprised at the ease with which they combined, draw them from the mass. This process, repeated so as to obtain a thread exceeding the length of the original fibres, would give the first woollen thread." Whether the importance of this first discovery vaguely dawned upon the mind of this hypothetical shepherd-boy or not cannot now be told; neither can it be known whether he carried his spinning operations beyond the first stretch or not. Only conjectures can be offered regarding its beginning, and these may be either far from or near the facts. But clearly a time would come when this would be done, and as a greater length of yarn was produced, to prevent its entanglement would involve winding it upon a twig. Here comes into view the beginning of the spindle, as yet, however, not yet used for its present purpose, or even such a purpose divined. But this in due course would grow out of the former. In order to prevent the unwinding of the yarn from the twig, in the event, say, of its falling from the hands of the spinner, the thread would be secured in a cleft made for the purpose at one end of the twig. Now further suppose, for there is nothing else available, that the spinner, after laboriously twining a length of yarn, instead of winding it upon the twig as usual, rises to his or her feet and allows the latter to dangle from the hand suspended by the length of yarn just spun, a new phenomenon occurs. The twig begins to revolve, slowly at first, but with increasing velocity, until suddenly, whilst the spinner is contemplating this vagary, the thread breaks, and the twig drops to the ground. The spinner then finds that all the twist has been taken out of the fibres. The rapidity with which the fibres would be untwisted, compared with the time it had taken the operator to twist them, could hardly fail to be recognised. Slowly it would break upon the understanding that if this revolving twig could thus take twist out by a reversion of its movement, it could be made to put it in.

"The mental suggestion would be acted upon, and the trial would succeed. It can easily be imagined how that spinner would exclaim 'Eureka!' 'Eureka!' that is, if Greek happened to be one of his accomplishments, which is not probable. This would be the first spinning spindle, that is, if our conjecture has been fortunate enough to hit the mark. All the steps between this rude discovery and the perfect form of the spinning machine with which the world is now acquainted consist of a series of improvements upon the original form."

The above is a very graphic and apt pen picture of the first woollen spinner and his surroundings, and from some such beginning as that outlined above originated the

distaff and spindle, ages upon ages before history can tell
us anything about spinning.   How many centuries elapsed
before the improvements had reached the perfection of the
distaff and spindle we know not; but this we know, that
in reaching that point the inventive genius of mankind
seemed to have exhausted itself in that direction, for

FIG. 49.—THE SPINSTER, DISTAFF AND SPINDLE.

thousands of years passed by before another forward
movement was effected, as we find that all the nations of
antiquity used the distaff and spindle.   Egyptians, Babylonians, Hebrews, and Phœnicians all used the distaff and
spindle, and, according to Wilkinson's *Manners and Customs
of the Ancient Egyptians*, the hieroglyphics of a word signi-
fying in Coptic "to twist" is always found over ancient

Egyptian representations of persons employed with the spindle, and as a curious testimony to the unchangeable character of Eastern customs, it is of the same form as that now in use in Egypt. Several of these spindles have been found at Thebes, of about 1 foot 3 inches in length. They are mostly of wood, and for the purpose of giving an impetus in twirling them the circular head is weighted with gypsum or some composition, and has a loop or cleft at the top for securing the thread after it is wound. The distaff was a short staff, to one end of which was attached the material to be spun, while the other end was held under the left arm, so as to leave the hands at liberty to draw the thread and manage the spindle. The Phœnicians, being the most skilful dyers of antiquity, dyed the wool before putting it into the hands of the spinner. Heeren, in his *Searches into the Politics, Intercourse, and Trade of the Principal Nations of Antiquity*, vol. i. p. 36, says : " It is to be regretted that history, which so celebrates the garments and woollens of the city of Tyre, has preserved us no distinct information respecting them." In Mitford's *Greece*, vol. i. p. 159, edition 1835, it is stated that "it was customary in the heroic age, as indeed at all times in Greece, for ladies of the highest rank to employ themselves in spinning and needlework, and in at least directing the business of the loom, which was carried on, as till lately in the Highlands of Scotland, and among the yeomanry in many parts of England, by every family, or its servants, for itself. It was praise equally for a slave and a princess to be skilful in works of this kind." The textile fabrics of both the Greeks and Romans were almost entirely of wool. They wore very little linen or cotton cloth, and the ancient nations adopted silk sparingly, as an exceedingly costly article of dress. Every considerable

house, especially in the country, contained a loom, together
with apparatus for preparing and spinning the wool, though
there were a class of persons called *textores* who manufactured for the public in the towns, but it was the general
custom for families to manufacture their own clothing, and
the dexterity acquired by long traditionary habit, and by
great delicacy of hand, will alone enable us to account for

Fig. 50.—The Hindoo Spinner.

the great beauty and fineness of their textile productions,
especially when we bear in mind the rude and simple
implements at their command.   Smith's *Dictionary of Greek
and Roman Antiquities* says : " The spindle was a stick 10 or
12 inches long, having at the top a slit or catch in which
the thread was fixed, so that the weight of the spindle
might continually carry down the thread as it was formed.
Its lower extremity was inserted into a small whorl (*ver-
ticellum*) made of wood, stone, or metal, the use of which

was to keep the spindle more steady, and to promote its rotation; for the spinner, who was commonly a female, every now and then twirled the spindle with her right hand, so as to twist the thread still more completely; and whenever, by its continual prolongation, it let down the spindle to the ground, she took it out of the slit, wound it upon the spindle, and having replaced it in the slit, drew out and twisted another length" (see Fig. 49, "The Spinster").

In the Introduction to this text-book reference was made to the extraordinary dexterity and skilful manipulation of the Indian spinster, and to what an amazing degree of fineness she could carry her thread, and there can be little doubt that the birthplace of the distaff and spindle was India. After long ages a knowledge of this mode of spinning moved slowly westward, through Persia and Arabia, and we learn incidentally that it existed in Egypt some two thousand five hundred years before the Christian era. Herodotus, one of our oldest historians, speaking of India in the fifth century B.C. says, "They (in India) possess likewise a kind of plant, which, instead of fruit, produces wool of a finer and better quality than sheep, and of this the Indians make their clothes." The industry moved slowly from one region to another, for it was on one side of the Mediterranean Sea for a period of thirteen hundred years before it crossed over into Greece. The spinner's art was early exercised on flax and cotton as well as upon wool. The finest flax was produced on the banks of the Nile in ancient days as well as now, though there is reason to believe that the fleece of the sheep was probably one of the first materials made into cloth; the manufacture of both linen and woollen existed in Greece in the days of Homer. The spinning by distaff and spindle was the only means of producing a textile thread through the long period of thousands of years; so long was this the only mode of spinning, that mankind seemed to have settled down to the conclusion that there could be no other, and the exact date at which the one-thread wheel was introduced is involved in some doubt and uncertainty, though it is comparatively a modern invention. The one-thread wheel enabled the spinner to produce seven times more yarn than by the distaff. There can be little doubt, however, that the crude old teak-wood

Fig. 51.— The One-Thread Wheel.

wheel of India is its progenitor. The one-thread wheel came into use in Europe at the latter end of the fifteenth or the beginning of the sixteenth century, and continued to be our most expeditious mode of spinning till the Hargreaves jenny appeared. Of course, in the homes of well-to-do people more elaborately constructed wheels were used, like Fig. 52. The jenny was soon followed by Crompton's mule, and the whole three modes of spinning were in active use for many years. The one-thread wheel can still be found in some of the remote districts of both England and Scotland. The Saxony wheel appeared about 1533, and was said to have been invented by a native of Brunswick, and was used chiefly for flax spinning, and but little used for wool.

With the advent of the Hargreaves jenny, machine spinning may be said to really commence. Though it is called a *hand jenny*, it was to all intents and purposes a machine, as its size was only limited by the ability of human strength to work it. The earliest mention that we have of Hargreaves is about the year 1762, when he is said to have been helping in the erection of a carding machine for Mr. Robert Peel of Blackburn, the founder of the Peel family. This carding machine was originally the production of Lewis Paul, the partner of John Wyatt of Birmingham, who undoubtedly invented spinning by rollers, and took a patent out for the invention in 1738, full particulars of which may be seen in Baines's *History of the Cotton Manufacture*. Wyatt made little headway with the invention, chiefly because he could not get the mechanical details fitted with such exactness as to enable the machine to work properly, and so the thing had to lie in the background till Arkwright heard of the principle of spinning by rollers, and grasped the idea at once, and made it the basis of his Water spinning frame. Arkwright had a most wonderful and extraordinary faculty of being able to see the worth and import of other people's original ideas, and being able to turn them to practical account, he could, as it were, instantly see the salient point of any new invention. At the break-up of Paul and Wyatt's establishment Paul's invention of the revolving cylinder for the carding of cotton was bought by a hatter of Leominster, and was applied to the carding of wool for hats. The machine was introduced into Lancashire about 1760, and attracted the notice of the first Robert Peel, who took his wife's brother, a Mr. Haworth, into partnership along with Mr.

William Yates of the Black Bull Inn, Blackburn (whose
daughter became the wife of the second Robert Peel, and
mother of the statesman), and hence was formed the noted

firm of Haworth, Peel, and Yates about 1750.   The first
Robert Peel was a man of keen practical insight, and tried
Paul's revolving cylinder carding machine, and found
Hargreaves to be a very useful man to carry out his

suggestions and put them into practical shape ; but even in Peel and Hargreaves's hands little progress was made till Arkwright's extraordinary faculty of putting into ship-shape the produce of other people's brains again came into play. Arkwright laid hold of the revolving cylinder, and used it effectually in the cotton manufacture. Kaye's invention of the fly-shuttle gave an impetus to weaving, and greatly increased the production of the loom, thereby causing the supply of yarn to fall far short of the demand, there being quite a scarcity between the years 1760 and 1767. Some half-dozen or more schemes were attempted to meet the difficulty that arose during the years named, but without much practical result. In the district round Blackburn Hargreaves heard much of this scarcity, being himself a weaver at Standhill, but being a ready-handed man, could do a little carpentry or mechanical work ; but the scarcity of yarn, and the consequent want of employment around him, exercised his thoughts very much, and fastened upon his attention to such a degree as to become the one absorbing subject of his study. Long and anxiously he turned the matter over in his mind, recurring again and again to the various points and difficulties, until at last, after long and strenuous thought, he saw his way to the construction of the spinning jenny in 1764 to 1767. With Wyatt's roller spinning idea it had not the slightest connection, the principle on which the jenny is based being quite opposite to Wyatt's principle. It was a purely original invention, and is, in fact, a machine application of the original principle of spinning, whereas Wyatt's rollers and Arkwright's water frame were not only new machines, but embodied a new principle of spinning altogether. Hargreaves's jenny produced yarn for the first time on a large scale on the old principle, whereas Wyatt and Arkwright's was an entirely new principle.

Hargreaves is supposed to have invented his jenny about 1764, and used it only to supply himself and family with weft, the warp at that time in the Blackburn neighbourhood being usually linen, as the district produced strong printing cloths. The invention had certainly been brought to completion in 1767, and the secret of the existence of the jenny could no longer be kept; the inventor's position, like that of Kaye, the inventor of the fly-shuttle, became one of extreme personal peril, as the working classes of Blackburn were not students of political economy in those days, any more than now. They only saw the immediate effect of increased production of yarn on wages, and not the ultimate effect of increased employment. A mob gathered, and broke into the inventor's house, destroyed his jenny, and gave him to understand that he could only remain in the neighbourhood at the risk of his life, so he had to fly to Nottingham. At the time of the invention of Hargreaves's jenny it took three to six spinners on the one-thread wheel to keep one weaver going, so that the jenny only came into being in response to an urgent demand. Meanwhile Arkwright had matured his roller spinning water frame, and enrolled the specification of his first patent in 1769, and, like his forerunners in that great era of invention, had been compelled to seek safety in flight. Without the pioneer Hargreaves there might have been no Arkwright, and there could have been no room for a Crompton. When Arkwright's first patent expired in 1784 there were in England 1,600,000 Hargreaves jenny spindles at work, and to Hargreaves belongs the honour of being the first inventor of a machine by which one individual could spin the fleecy and fibrous substances of the animal and vegetable kingdoms into a plurality of threads at the same time, and by one operation.

Within two years of the invention of the Hargreaves jenny we had Arkwright's water frame, and we reach a very important period in the history of modern spinning at which the road parts into two. The perfecting of Wyatt's roller spinning frame by Arkwright introduced into the textile industries a completely new mode of spinning that had not to any degree been in use before. *All our textile threads, up to the point at which the Arkwright frame appears, had been produced on one and the same principle, the original principle of drawing and twisting at one and the same time, of distaff and spindle and the one-thread wheel; this was continued and extended in the Hargreaves jenny.* The Arkwright frame brings into view a new principle of spinning, and converts the spinning of cotton, worsted, flax, hemp, jute, mohair, alpaca, spun silk, and indeed every other non-covering yarn that can be named, whether composed of vegetable or animal material, into two distinct and separate simple processes which follow each other continuously. First the material is drawn out to the requisite degree of fineness by a series of rollers placed one behind another, each roller running a little quicker than the one immediately behind it; the second simple operation, which immediately follows the first, is that the half-formed thread as it emerges from the front roller is attached to the spindle and twined or twisted to the requisite degree to bear the tension of weaving and all the after processes. The drawing of the thread by rollers is one process, and the twisting of the thread when it reaches the spindle is another, and that which had hitherto been one compound process, intermittently carried out, was by the Arkwright frame divided into two simple processes, each distinct and separate from the other, but following each other, and which can follow each other *continuously.*

In the Arkwright frame, in 1769, continuous spinning
appears for the first time in history in a practicable
working form.   Hitherto the process of drawing and
twining had not been two separate and distinct pro-

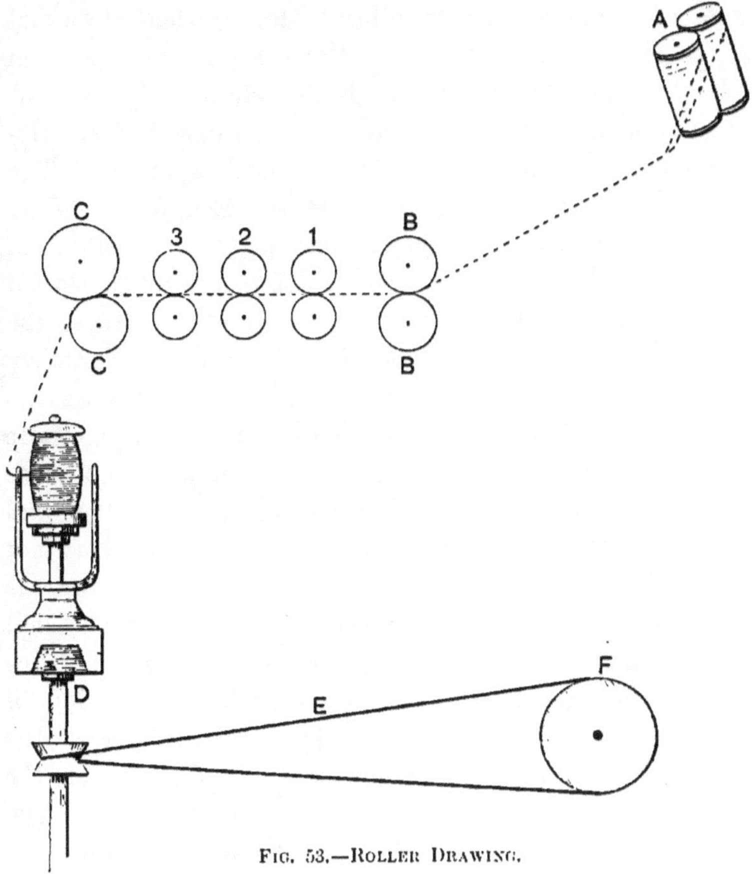

Fig. 53.—Roller Drawing.

cesses, separately conducted one after the other, but
the two had been carried on together as one compound
process; so that the Arkwright frame introduced quite *a
new principle of spinning* into the production of textile
threads, *and divides yarns into two distinct classes.*   One class

comprises worsted, cotton, flax, hemp, jute, mohair, alpaca, spun silk, in fact all the *wiry, bare*, non-covering yarns, of whatever material composed; and the other class comprises woollen and all its allied yarns, whether composed of wool in combination with other material, whether animal or vegetable, that are not roller-drawn, but spun on the old principle of the one-thread wheel and the Hargreaves jenny. We have now two distinct and clearly-defined methods of drawing in our textile spinning. The one method of drawing *by rollers* (see Fig. 53) converts wool into a worsted thread, and the other method of drawing *by the spindle* converts wool into a woollen thread (see Fig. 51); the difference in drawing constitutes the dividing line between worsted and woollen. Formerly only long wool used to be spun into worsted, but now the improved combing machinery enables the spinner to use almost any kind of wool at will. Identically the same kind of wool is frequently used for both woollen and worsted. Combed wool and uncombed wool does not make the difference between woollen and worsted, as carpet worsteds are not combed, neither are fingering and many other kinds of hosiery worsteds. The difference between worsted and woollen is neither in the kind of wool nor in the combing; the line of division is in the drawing. The modern worsted thread is drawn on the new principle of spinning-drawing *by rollers* that is embodied in the Arkwright frame; the woollen thread is drawn on the old principle of spinning-drawing *by the spindle*, and is embodied in the Hargreaves jenny and the old one-thread wheel. Prior to Arkwright's frame, introduced in 1769, our textile threads were all drawn on one principle, but since 1769 we have had two principles of drawing at work, and consequently two systems of spinning in our

textile industries. In the new principle of spinning the
spindle has nothing whatever to do with the drawing; it is
all done by rollers. In the old principle of spinning the
rollers have nothing whatever to do with the drawing; it is
all done by the spindle. If we are to make any satisfactory
progress in the elementary knowledge of the structure and
uses of textile threads we must endeavour to fathom the idea

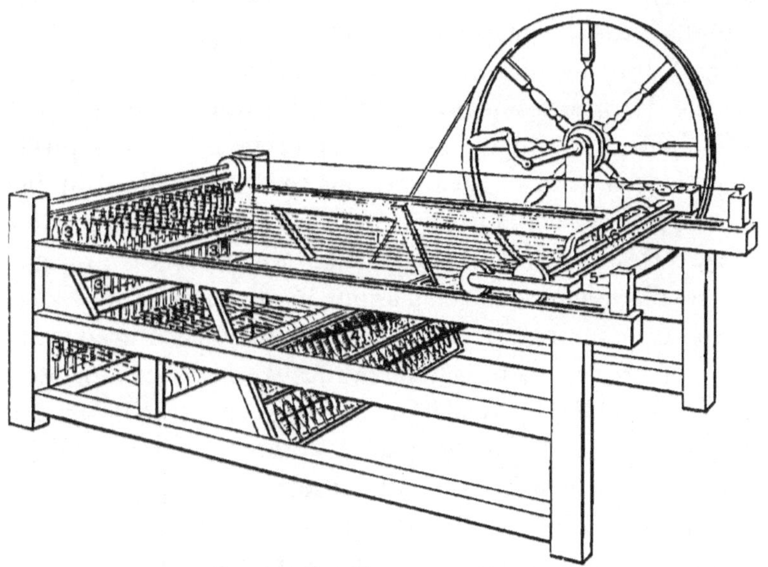

Fig. 54.—The Hargreaves Jenny.

here at its very fountain-head. On one side we have arrayed
the roller-drawn yarns, a very large class, say 70 per cent of
our textile yarn; and on the other side the spindle-drawn
yarns, not so many in number as the other class, but of very
ancient date, or rather of prehistoric date as to the origin of
the principle of their structure. The distinction between the
two classes of textile yarns is sharply defined at the fountain-head, where the roller yarns branch off. This class is
roller-drawn, and the other class is *spindle-drawn*. The

student will learn farther on what is involved in these two distinctions, as they will become the subject of our debate; and if we can lay in, as it were, a little alphabetical insight at this point in passing, it will be doubly useful later on, as a very little well-grounded elementary knowledge goes a long way towards mastering the difficulties of the subject, and the simple elementary distinction here is *roller-drawn* and *spindle-drawn*. A great deal will arise out of this simple distinction of roller-drawing as exemplified in the Arkwright drawing frame, and spindle-drawing as exemplified in the Hargreaves jenny (Fig. 54). From these two root principles of spinning arise all our textile tissues or yarns, and the character of each yarn has to be determined by the source from which it has sprung--the method of its production. How much is implied in this we must leave for our future discussion to develop as our inquiry proceeds; suffice it to say here that in order that we may name aright any tissue that may be placed before us, we have first to ascertain the mode by which it has been produced—is it a roller-drawn or a spindle tissue? When this is ascertained our name for it will soon be forthcoming.

The invention of the Hargreaves spinning jenny was followed after two years by the invention of the Crompton spinning mule, one of the most efficient machines, even to this day, that can be found in the whole range of our textile industries. In 1779 Crompton had completed his spinning mule; he took the upright spindle frame of the Hargreaves jenny and placed it on a movable carriage, and then took the rollers of the Arkwright water frame and placed them in the position from which he had removed the spindle frame of the jenny. By taking the best parts from each machine, and combining those parts, he succeeded in producing one of the best machines of modern times. The roving or the slubbing, as the case might be, he placed in a creel behind the spindles, instead of in front, as in the jenny, to a position, in fact, behind the rollers. We have just said that the rollers of the water frame were placed in the position of the stationary spindle frame of the Hargreaves jenny, only a little higher, and consisted precisely of the same series of rollers, one behind the other, right back to the roving creel; the row of rollers farthest back against the roving creel move the slowest, the row immediately in front of the back ones move at a quicker speed, the row in front again move quicker still, and the front row near the spindle quickest of all; so that the new principle of doing the drawing by the rollers was carried out as completely in the Crompton mule as in the Arkwright water frame, with the additional advantage of the spindle draft. The mule is a wonderful combination of rich spinning properties to those who will take the trouble to study them. Crompton says in a letter to a friend : " In regard to the mule, the date of its being completed was in the year 1779. At the end of the following year I was under the necessity of making it public or destroying it, and to destroy it was too painful a task, having been four and a half years at least, wherein every moment of time and power of mind, as well as expense, which my other employment would permit were devoted to this one end, the having good yarn to weave, so that to destroy it I could not." Being of a retiring and unambitious disposition, he took out no patent, and only regretted that public curiosity would not allow him " to enjoy his little invention to himself in his garret," and so earn, by his own manual labour, undisturbed, the fruits of his ingenuity and perseverance [1]

----

[1] See Baines's *History of the Cotton Manufacture.*

(see Fig. 55). Whatever Arkwright could do on his water frame the Crompton mule could always do, and something besides, which Arkwright could not do. However Arkwright improved his yarns by improved carding and drawing, the mule could always step in and give a finishing touch that was utterly beyond Arkwright's reach. As it was at first, so it is now; the mule as a cotton-spinning machine can do anything the ring frame or any other frame can do, and do it better; it combines in one machine both the new and the old principles of spinning, and can combine both roller draft and spindle draft in one and the same yarn or thread. The mule in its native element—the cotton industry—rarely, if ever, dispenses wholly with the spindle draft, but the mule applied to woollen is entirely a spindle-drawing machine.

Fig. 55 is a representation of the Crompton mule, and is taken from a mule made and worked upon by Samuel Crompton of Bolton, now in the Chadwick Museum of that town, and is the property of Messrs. Dobson and Barlow, Limited, of Bolton. In a paper on "Some Difficulties in Cotton Spinning," by B. A. Dobson, C.E., M.I.M.E., read at the Bolton Technical School on 10th January 1893, in speaking of the above mule, Mr. Dobson says, "You may see to-day in the Merchall Museum in this town an old wooden hand mule which was made and worked by Crompton, the inventor of the mule. It is by no means the first or the only one he constructed, but is one upon which he worked and earned his living at the beginning of this century. This rudely-constructed machine contains every motion, or at least an equivalent of every motion, of the most complicated and perfected self-actor of to-day. The machine was driven by power, and the driving of the carriage out, and the twisting and stretching and drawing, were all done by power. The backing-off and running-up, equivalent to the winding-on in the hand-spinning wheel, were accomplished by manual power, guided by intelligence. But the after-stretch, which means that the rollers stopped before the carriage had completed its journey, and consequently, whilst twisting, the yarn was still stretched by the

Fig. 55.—THE CROMPTON MULE.

carriage spindles; the extra twist motion, which means that when the roller and carriage were stopped, the spindle was still turning on to put in the proper amount of twist; and the receding motion, which, whilst this last-named operation was going on, allowed the carriage to approach a little towards the roller—the necessity for all these motions was well understood and allowed for. In 1878 my firm showed this mule at the Paris Exhibition. The mule was fitted up as complete as possible, and before its departure I spun some No. 80's upon it. I therefore speak with a knowledge of the fact. I do not, however, profess to be a hand mule spinner, and fear that the cop I produced might have excited the derision of an old spinner. Still, it was yarn and a cop. Now, I do not intend for a moment going into mechanical details, because, although to trace the evolutions of the mule would be a task worthy of any one, it would require more time than a lecture affords. But the machine consists of an agglomeration of details all worthy of discussion, both instructive and interesting. So far nobody has written what might be termed 'The Romance of the Mule.' Some of the best brains of Lancashire have been racked, and are still being racked, with a desire to further improve this machine."

In the mule the spindle-bearing carriage moves backward and. forward, or in and out, with every stretch of yarn, just in the same manner as the draft carriage bearing the clove or clasp slide did in the jenny. In the mule applied to woollen no use whatever is made of the series of drawing rollers ; they are dispensed with altogether, as there is no roller-drawing in the properly constructed woollen thread. The only use there is for them is that the use of the front rollers is found to be the most convenient and uniform method of getting in the roving or sliver, and to hold the same fast while the spindle carriage recedes and draws the thread. When the rollers begin to wind in the sliver or roving, the sliver attached to the spindle is borne away from the rollers, at the same rate as the rollers deliver it, twine or twist being put in by the spindles all the time ; and when about half the length of the stretch of the mule has been wound in, the rollers stop and the spindles commence drawing out the thread, revolving and receding all the time, thereby forming the thread. The thread thus produced is identical in character with that produced by the one-thread wheel and the Hargreaves jenny. The rollers have done nothing but wind in the sliver, and then stopped and remained perfectly at rest during the forming of the thread by the revolving and receding spindle. We shall have to refer to this point again and again as our inquiry and discussion proceeds.

The mule during the first twenty years of its existence was operated or worked by manual labour, like the jenny. A Mr. Kelly of Glasgow, or Mr. David Dale of Lanark, was the first to apply steam or water-power to the Crompton spinning mule by means of a loose pulley, into which a catch worked that could be thrown in and out of gear at pleasure. This primitive mode of action was speedily superseded by the fast and loose pulley, and an additional rim was brought into use to increase the speed; an additional fast pulley was fixed beyond the other fast pulley. To throw the strap off the fast pulley at the proper time when the stretch or draw of yarn was completed, a worm was placed on the extreme end of the rim shaft, outside the framework of the head-stock, and into this worm a wheel worked which registered the revolutions of the rim shaft, and at the proper time, by raising a catch, threw the strap back again to the loose pulley by means of a long swivelling lever, to one end of which was attached a cord and weight, and on the other end of the lever was the loop or guide through which the strap ran. French's *Life of Crompton* says David Dale of Lanark was the first to apply water-power to the mule. Very soon after the mule was introduced Mr. Wright of Manchester removed the head-stock from one end to the middle of the carriage. I well remember that the first mules upon which I was employed were a pair, of 100 spindles each, very old machines, with the head-stocks at the ends of the carriages, and as a consequence the mules were what were termed " right " and " left " handed, and great used to be the fun when a strange spinner happened to come into the room and tried " to put up " the mules, for in " putting up " the " uncanny things " the spinner had to use his left hand in place of his right, and his right hand in place of his left, and to transpose the position of his body every minute, first right foot first, then left foot first ; few spinners could change front so rapidly and so often. The removal of the head-stock by Mr. Wright from the end to the middle of the spindle carriage led at once to the carriage being lengthened out to double its previous length ; with this increase in its length another difficulty came to the front. From this cause it was found to be very much wanting in rigidity, and this led to a different arrangement of the cross-stays in the building of the carriage, and to two lengths of strong wire being strung from end to end of it, the wires being crossed from front to back of its under part, and then from back to front every two yards, crossed continuously from end to end, and when tightly braced, were found to give sufficient rigidity to make the carriage practically workable without materially increasing its weight in build. Then a further difficulty arose in getting the carriage to move parallel with the roller beam, and this was ultimately overcome by the invention, or rather the new application of the squaring bands. With the removal of the head-stock to the middle, or nearly to the middle of the carriage, the " right and left " hand mules disappeared, as both mules of a pair were by this change made right-handed, for the head-stock on one side was a little past the middle and the head-stock on the other was a little short of the middle, so that the spinner had simply to turn round from one mule to the other and take hold of the wheel with his right hand and the faller with his left hand—all mules becoming right-handed by the change. With this great increase in the size of the mule its management required corresponding care and skill.

The next move in the way of advance was Eaton's automatic copping motion, in 1818, which was the initiatory move towards making the mule self-acting. Many persons tried their "prentice hands" at accomplishing this object, but not much progress was made beyond the point reached by Eaton until Richard Roberts of Manchester took the matter in hand. Previous to the mule being made self-acting there was quite a craze for running out the machine in length ; people seemed as though they would never be satisfied with size. At one time I was employed upon a pair 540 spindles in length, or 1080 spindles for the pair of $2\frac{1}{2}$-inch pitch per spindle, and then upon a pair 600 spindles in length, making a total of 1200 for the pair. This was a great length for woollen mules. The craze, however, abated when the mule was made self-acting, and about 300 spindles each mule is now most common in the woollen manufacture.

In 1786 another machine, which may be said to be an off-shoot of the principle of the mule, was invented by some one near Manchester, and was called a *billy*. It came rapidly into use, and was used as a machine in the cotton manufacture for preparing the slubbing for the water frames and the mules. The billy was used in the woollen manufacture long before the mule. There was such an urgent need for the Hargreaves jenny, that when it appeared it seems to have been eagerly laid hold of and put into use all along the valleys and hillsides of the Yorkshire woollen districts, being set up in the chambers of the houses, or in any room that could be spared for the purpose about the buildings of these domestic manufacturers. The billy betook itself directly to the mill, and was, of course, always from the first a mill machine, having to stand close beside the carding machine, from which it received the cardrolls of wool as they dropped out. The billy, like the jenny and the mule, was in its early days a manual machine, and when a young man I used to see almost daily the slubber turning his billy by hand, amid the whirl and noise of wheels and the swing of straps. How strange this sounds now! but our forefathers were never in a hurry to apply steam to anything that they could do by their own hand labour. The billy took the rolls of carded wool or cotton, which were pieced end to end by children on the sloping board behind the machine, and formed them into slubbing or roving, as the case might be, according as the machine was applied to woollen or cotton (see Fig. 56). Nearly all the machines for the early processes of the woollen manufacture were first invented for, and used in the cotton industry; but the billy looked so well adapted to its work, that very few persons knew or could have imagined that it was invented for or applied to any other industry than woollen. Though uncouth in appearance and ungainly in its action, the billy in its day has been a very useful and effective machine, more especially after the piecing machine was brought out and child labour dispensed with in the working; and this much further may be noticed respecting it, that during its reign the woollen manufacture attained to a degree of excellence that it has failed to maintain, or reach, since the billy has been abandoned. The reason for the decline in the woollen manufacture consequent upon its disuse we shall come across as our discussion advances. The billy had its place, and a very important

place, too, which has never since been filled, and did important work by carrying forward the first operation of spinning,
not only in the woollen, but also in the cotton.   The piecing
machine got rid of hand-piecing with all its faults, and

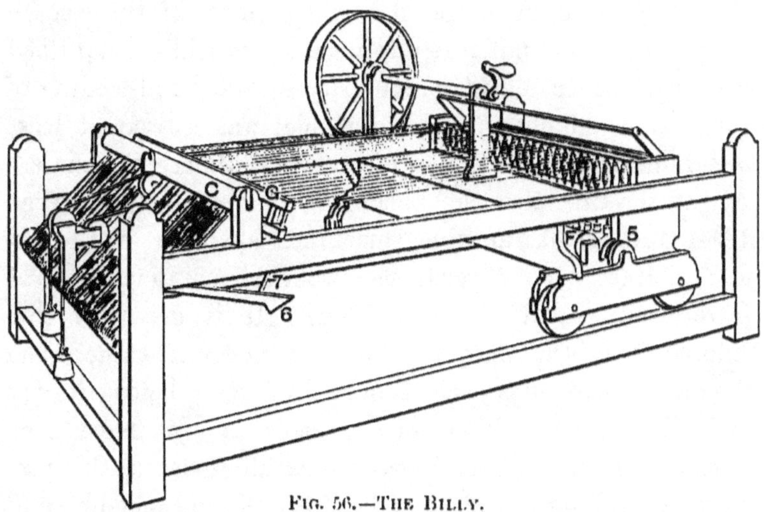

FIG. 56.—THE BILLY.

The billy was invented for the cotton manufacture, and the above illustration
copied from *The Illustrated Gallery of Arts* shows the machine in its early form
with the wheel fixed to the end of the spindle carriage, and with a supporting
pillar standing on the middle of the carriage to bear up the end of the shaft carrying the handle.   The billy as shown above is slubbing cotton, and it was formerly
called a slubbing jenny, being, in fact, an adaptation of the Hargreaves jenny.
There is at the end of the machine a sloping board, on which the rolls of wool or
cardings, about a yard in length, were placed, side by side, by children called
"piceners," and it was the office of the "picener" to place new cardings on the
board, and to squeeze and rub the ends to the previous cardings, so as to form
continuous slivers.   The slivers were carried up by the endless apron and under
the roller C, and then between two bars called the slide, like the "clove" in the
jenny, the upper bar G, which was raised to allow of their passage to the spindles.
When a sufficient length had been drawn through, the upper bar was let fall,
and the portion of sliver drawn in was attenuated or drawn out and then wound on
the spindles.   When the carriage returned home a wheel 5 ran under the lever 6,
which raised the sliding rod 7 upon which the bar G was mounted, thus opening
the clasp or slide for another length of cardings.

henceforward there was no more "rackin' the rowins," as
the Scotch called the stretching of the cardings by the
children, and the feeding machine got rid of the inequalities

of hand-feeding, and was another instance of the adaptation of cotton appliances to woollen purposes. The wool was taken off at the side of the breaker scribbler in a sliver, and fed upon the intermediate scribbler, either in the form of a bank-feed, or in the flat ribbon form of the Scotch feed. In some mills Apperly and Clissold's harp feed was used, so that any irregularity in the hand-feeding of the first machine or breaker scribbler was crossed at least one hundredfold, and the irregularity almost eliminated. About the time that the piecing was invented, or between 1830 and 1840, another machine, alien in its origin and character as regards the woollen manufacture, was introduced, called the condenser. It is generally considered that the first idea of the machine came from America; the idea was thought of long before it was embodied into working form. It is said to have been named the "condenser" because it abbreviated the processes by taking the place of both the piecing machine and the billy. It fatally happened that it *condensed* at a point where *expansion was very much needed*, and just at a period, too, in the circumstances of the woollen manufacture when such an expansion could have been most easily carried out. It not only condensed at a point where expansion was much needed, but barred the way and made the place barren, so that no other machine could do the work so essential at this stage of the process.

While Richard Roberts was improving the mule to the utmost, and making it self-acting by the addition of one motion after another in rapid succession, the "condenser," which was to *dispense altogether with the greatest portion of the spinning*, was being introduced. The introduction of the condenser displaced nineteen-twentieths of the spinning at a time when the woollen industry was actually hungering for an in-

crease of the spinning-drawing, in order to more fully develop
the structure of the woollen thread, and thereby open out a
possibility for the advance and further development of the
manufacture.   The sister wool industry, the worsted, was
just at this time extending the drawings and recrossings,
and further developing the worsted thread and making it
suitable for a more extended use; while the woollen industry was curtailing the drawing and making the woollen
thread unfit for the uses to which it had been applied
for centuries.   Surely such an ill-considered and unwise
step was never taken before in connection with any
industry, ancient or modern, as dispensing with so large a
percentage of spinning-drawing, when it ought to have
been increased.   If the importance of woollen spinning
in the manufacture had been clearly understood, no *con-
densation* would have been allowed, more especially at the
very period in its history when *expansion* was so urgently
needed, in the best interests of this ancient industry, to fit
it for stepping forward and extending the field of its manufactures.   It was not intelligence but ignorance that threw
away 95 per cent of our woollen-drawing to make way for
the condenser.   The worsted people have extended and
expanded their drawing, and perfected their thread, while
the woollen people, by their lack of technical and scientific
knowledge, *and their folly*, have almost ruined the woollen
industry.   While the worsted people have expanded and
extended their crossings and redrawings into thousands
of millions, we in the woollen have "condensed" or
reduced the amount of our drawing, which was before insufficient.   It is by drawing on the woollen principle alone
that a woollen thread can be formed, and during the last
half-century the amount of drawing bestowed upon the
woollen thread has been continually reduced, as a thing to

be got rid of, and, as a consequence, the woollen manufacture has been reduced to the contemptible position of being able only to produce "fancy woollen goods," and that, too, with only indifferent success.

We cannot produce the fine super west of England broadcloth, as in days gone by, from our present condenser thread; this is an example of a full milled cloth. Then, on the other hand, we cannot produce a Bannockburn tweed fit to be shown, and this is a cloth at the other extreme of our manufacture, and requires no milling whatever. Our present condenser thread is utterly unworthy of its name; it is a tissue of fibres that are not half formed into a thread, indeed, it has not received a tithe of the drawing necessary to constitute a thread; it is only a crude conglomeration of fibres "without form and void." Why should the woollen manufacturer forget, or disregard, what he ought to know, namely, the principle on which the woollen thread is formed? Are scientific principle and technical knowledge nothing to the woollen industry? Our worsted friends on the other side of the wool industry do not so listlessly regard things that are vital to their success, but eagerly seek at every turn to establish their manufacture on a clear and scientific basis; even in the face of the keenest competition they have pushed improvement to its utmost limit. Get the desired result first, and then talk of the cost of production afterwards in its proper place.

The woollen manufacture in its early processes, instead of having kept abreast of the times, is less scientific in its procedure than it was fifty years ago. The thread obtained by one drawing from the condenser is a miserable substitute for a real woollen thread. In the old billy days the manufacture was much better served, for then there were always two, and sometimes three drawings, when roving was resorted to ; but now in the majority of cases we do not get one perfect drawing, and one drawing fails to produce a complete woollen thread. The fact is that the woollen thread, to our shame be it stated, is now very much less completely formed in our mills than it was fifty years ago, and, as a consequence, goods that used to be produced have had to be abandoned, and the goods that can be produced from it now are *of more limited range*, in consequence of the *more limited capability* in the character of the yarn ; and had it not been for the almost endless variety of patterns brought out in connection with the fancy woollens, the manufacture would long ago have been at a deadlock. If, for instance, a twenty-four skeins yarn is now required to be produced, it is quite common to condense the wool to sixteen skeins, leaving only a miserable half unit for drawing, out of which to build up a thread ; whereas in the old billy days the cardings might have been seen dropping out at the end of the carding machine at about $2\frac{1}{2}$ or $2\frac{1}{3}$ skeins in counts or thickness, and were then taken to the billy and drawn out into twelve skeins slubbing, or to five times its original length. The twelve skeins slubbing would go to the mule and be drawn into a twenty-four skeins yarn, so that the $2\frac{1}{2}$ skeins carding gets drawn to about ten times its original length, or in other words, it is a yarn or thread built up by receiving ten units of drawing. The unit of drawing here means that when one yard is drawn out into two, or two into four, four into eight, eight into sixteen, sixteen into thirty-two, each instance constitutes a unit of drawing ; so that the unit of drawing always means an extension of the thread by drawing into double its original length, and ten times its original length means ten units. The student will please bear this in mind, so that what may be said afterwards, when units of drawing are mentioned, may be intelligible. Note very carefully the difference in the comparison that follows. The sixteen skeins condenser sliver does not get drawn so much as to double its original length; it only receives one-half of that amount of drawing. To have received a unit of drawing, the sixteen skeins sliver would have required to be drawn to thirty-two skeins, whereas it was only extended to twenty-four skeins, or half a unit, so that the billy yarn gets twenty times as much drawing as the condenser yarn in this instance, and as the thread is formed or constructed by drawing, *wholly and solely by drawing only*, it does not take the wisdom of Solomon to find out which is the better thread, nor the insight and foresight of a prophet to foretell the disastrous effects of using such an ill-formed tissue as that from the condenser, *formed as above*, where a pure and properly built up woollen yarn is required, or even in place of the old billy yarn. We are not sure that some wiseacre will not, before the end of another year, put narrower rings upon the condenser, or even with the present rings condense the sliver to the entire length of twenty-four skeins, then carry it to the mule, and simply twine it, and forthwith take out a patent for the achievement, and then the insanity of the condenser craze would be complete, having done away with spinning altogether. We are already within half a unit of such a result; why hesitate about that narrow margin?

It cannot be too distinctly borne in mind by the student that in forming a thread there is not a particle of anything done till the spinning stage is reached; the carding and condensing have nothing whatever to do with forming the thread beyond leading up to the spinning stage. The thread is formed wholly and solely by the drawing of the spinning, and unless the spinning shall have conceded to it its place and range, we must expect the woollen manufacture to continue to decline.

While other textile industries have mostly a brilliant record to show of advances made during the last fifty years, we in the woollen manufacture have nothing to show, except it be "an advance backwards" from ten units of drawing down to a wretched *half unit*, 95 per cent of our former drawing trampled under foot, and as a natural consequence, our spinning has deteriorated, the woollen manufacture has had to beat a retreat all along the line, and other industries have occupied its place. The worsted coating has taken the place of the old west of England superfine black broad, which can no longer be produced in its old strength of fabric, fulness of face, lustre, and suppleness under the *régime* of the condenser. The worsted trousering is driving the woollen trousering out of the market, and unless the spinning can be speedily restored and given back to the woollen thread, we must be prepared to see the woollen trade decline still further, and *more perfectly tissued fabrics* take its place. The woollen thread *was never as badly formed and as imperfect at any period of its history as it is now, in the last decade of this nineteenth century of vaunted light and knowledge!!* In ancient times when spun on the distaff and spindle the woollen thread had more structure or build in it than it has now, when spun direct from the condenser, as it got much more drawing than the condenser thread gets in one drawing. When in olden times it was spun on the one-thread wheel (the bobbin wheel of to-day) the structure was much superior to our present condensed thread, as it got several units of drawing, whereas our present condensed thread very rarely gets so much as one unit of drawing,

much oftener only half a unit.   Then, a little nearer to
our own times, it may be said that the old *hand jenny*
advanced the drawing up to ten and sometimes to twelve
units, in conjunction with the billy. This is the highest
point that drawing has reached in the woollen manufacture,
as the mule did not increase the drawing, but simply took
up the lines of the hand jenny in respect to drawing ; the
mule only performed its work with greater mechanical
exactitude and with great increase of production.   We
cannot conceive a more painful and glaring instance of the
need of technical schools and the wider diffusion of technical
and scientific knowledge than that which is shown by the
present position and recent history of the spinning and
thread structure of the oldest of our textile industries.

Any kind of partially formed trumpery tissue has in our
day to do duty as a woollen thread, to the very serious
detriment of the manufacture, while at the same time the
worsted, cotton, and kindred industries, by drawing and
re-drawing, are carrying the structure of their threads to
a very high degree of perfection. Take the following
instance of the crossings and re-drawings in combing
and spinning a fine quality of 48s. worsted in the Bradford
district :

24 card slivers up at preparing gill.
 8 from preparing gill up at backwash.

192
 3 from backwash up at gill for comb.

576
 4 slivers drawn off comb.

2304
 20 comb slivers up at 1st finishing gill.

46,080
 4 balls from 1st up at 2nd finishing gill.

184,320
 6 tops up at 1st drawing gill.

1,105,920
 4 slivers from 1st up at 2nd drawing gill.

4,423,680
 4 slivers from 2nd up at 3rd (spindle) gill.

17,694,720
 4 ends from spindle gill at 1st drawing frame.

70,778,880
 4 ends from 1st drawing frame up at 2nd drawing frame.

283,115,520
 3 ends from 2nd drawing frame up at 3rd drawing frame.

849,346,560
 3 ends from 3rd drawing frame up at 4th drawing frame.

2,548,039,680
 2 ends from 4th drawing up at reducing frame.

5,096,079,360
 2 ends from reducing frame up at roving frame.

10,192,158,720

The roving is now ready for the finishing spinning, and
is either taken to the worsted spinning frame or to the
worsted spinning mule, and gets as much further drawing
as the condenser thread gets for its entire drawing. Need

we be surprised that the worsted thread is a success in almost every fabric in which it is used, or that the woollen thread, as at present constructed, is either an absolute or a comparative failure in every fabric in which it is used? and were it not for fancy cloths we should speedily come to an absolute deadlock. Do our worsted friends use all these elaborate appliances for the obtaining of a perfect worsted thread for nothing in these times of fierce competition, or are they not applied to obtain certain results; and if all these means are taken to produce a worsted thread, running into thousands of millions of drawings and crossings, will it never dawn upon the manufacturers in the woollen industry that perhaps some system of drawing is necessary in the woollen? All this elaborate drawing is resorted to in the worsted in order to lay all the fibres parallel, and to produce a thread as smooth and clear as drawn wire; and is no corresponding effort to be made in the woollen? A recent writer says "the fibres of wool lie in all possible directions in the thread." This is a grave mistake in regard to the woollen thread. The fibres of wool in the woollen thread lie only in *one certain specific direction*, not "in all possible directions," and the scientific principle of its structure is as clear and sharply defined as that of the worsted or any other textile thread in existence. There is one direction, and only one direction, in which the fibres can lie in a woollen thread; its structure is as absolute and scientific as that of worsted. It must no longer be supposed that any conglomeration of fibres gathered together higgledy-piggledy, by any sort of haphazard process, ought to be called a yarn — it is only so much spoiled material; a condenser sliver, girded together by twine, like a farmer's hay rope, is not a thread; it lacks all the structure of a thread, and has no rightful claim to any such name.

What the structure of the woollen thread is it will now be our endeavour to explain as the discussion proceeds. From this point forward the student will need to take special note of what is said as to the structure of the woollen thread, so as to grasp the principle involved, and so be able to follow intelligently the discussion onwards to its close. Our forefathers were not those absolute dunces that we too often take them to have been, for they handed down to us a woollen thread with twenty times as much drawing in it as the present condenser one, and unworthily have we valued and cherished our inheritance. The very least that the present generation ought to have done was to have added at least five or more units to the drawing, instead of letting the drawing run down to the miserable fraction of a unit. But ought so sweeping a statement to be made as that the old billy thread got twenty times as much drawing as the present condenser thread? Can such a statement be verified? Well, it is in the main already verified by the particulars stated as to the thickness of the old billy cardings. But how is this figure of $2\frac{1}{2}$ skeins in the thickness of cardings arrived at in the instance given respecting the twenty - four skeins yarn? Any elderly or even a middle-aged person, acquainted with the customary routine of a woollen mill before the disappearance of the billy, knows that the work was given out to card and slub at a certain price per wartern of 6 lbs. (more will be said in explanation of the wartern later on), and the putting-out price for a twenty-four skeins yarn would be about 2s. per wartern. The figure would be deviated from more or less in different districts, but it is sufficiently near the mark to serve as an illustration. Each individual mill had its scale to card to, and could state at once the number of cardings per half ounce that the work would be carried to as soon as the price was named. We will say, then, that our twenty-four skeins work was sent into the mill at 2s. per wartern, and that the mill carding scale for such a price was sixteen cardings or rolls of carded wool per half ounce; the reckoning would stand something like the following, viz. :—

```
        42 inches width on the wire of
              carding machine.
         1 inch to deduct for overlap in
        —       piecing cardings end to end.
        41
        32 cardings per ounce or 16 per
        ——      half ounce.
        82
       123
       ————
      1312
        16 ounces per lb.
       ————
      7872
      1312
     —————
     20992
         6 lbs. per wartern.
     —————
    125952
```

```
Divide by 36 or 6 × 6        6)125952
                             ————————
To get into yards            6)20992
                             ————————
Divide by yards in a skein    ) 3498 ( 2⅓ nearly of skeins per wartern.
    = 1520  .   .   .    .  .  ) 3040
                             ————————
                                458
```

Any one acquainted with the old routine can make a similar calculation for himself; it may differ slightly from the above, as different districts differ very considerably in prices and in scales of carding for given prices. The statement of $2\frac{1}{2}$ skeins is within bounds, and is rather under than overstated. We think that the statement of $2\frac{1}{3}$ to $2\frac{1}{2}$ skeins for the thickness of the cardings, in the old billy twenty - four skeins yarn, cannot be successfully disputed or set aside, and that the statement holds good that the old billy yarn got twenty times as much drawing as the condenser yarn of the present day.

The two wool industries traverse the whole field of modern spinning. The worsted embodies the new principle of spinning that was inaugurated by the Arkwright water frame, that is, the principle of *continuous* spinning, the drawing of which is done by rollers, and embraces the entire class of the clear, bare, wiry, non-covering yarns, say something like 70 per cent of the entire production of textile yarns, so rapidly has the principle of continuous spinning spread during the hundred and twenty-three years of its existence. The woollen, on the other hand, embodies the old principle of spinning, the *intermittent*, as distinguished from the continuous, the drawing of which is done by the spindle, and not by rollers. No roller-drawing whatever enters into the structure of the woollen thread ; it is purely a production of spindle-drawing. Worsted is spun on the *new* principle, and woollen on the *old*. The two threads or yarns have nothing in common, only the wool from which they are both spun. It is the two modes of drawing tending in quite opposite directions that bring out results so widely differing in character. And the fact that they are both from wool, and in many cases from identically the same wool, leaves us no alternative but to conclude, other considerations apart, that the difference in the two products arises from the two essentially different modes of treating the same wool. The principle of the new mode of drawing by rollers lays the fibres parallel, and that of the old mode of spinning lays the fibres in longitudinal waves, or corkscrew form, with each end of the fibre fringing out of the core of the thread. The woollen is a fringy *covering* thread, *not a bare, wiry one*, like the worsted, alpaca, cotton, flax, etc. The draft of the woollen thread can be more uniformly conducted in our modern mills than at the time of the old hand jenny and other hand processes. The wool leaves the last carding machine in our modern mills, and is received by the condenser, which in turn converts it into small sliver, or in other words, into light, filmy tissue, technically called sliver. Unlike the worsted, there is no interruption to take the sliver to the combing machine to have the noils taken out, but it goes forward bodily towards the spinning. The sliver on leaving the condenser is wound upon long bobbins, or spools, or reels, or whatever other name different districts choose to call the articles employed, and then carried to the spinning mule; where we must slacken pace and linger awhile, for here centres the gist of the whole matter of the structure of the woollen thread. Spinning here is not like spinning in worsted—a series of simple processes continuously going on, and succeeding each other—but is one compound process intermittently carried on. Let us explain, even at the risk of being tedious. The process is not, as in the instances of worsted and cotton, a continuous draft or drawing of sliver by rollers. There is no drawing in woollen by rollers; only one set of rollers is employed in the woollen mule, and the set that is employed is not for draft, but only to wind in the sliver, and hold it fast in its grip, the same as the slide in the billy, and the same as the clasp slide in the old hand jenny. The draft is done wholly and solely by the spindle, whereas the spindle has nothing whatever to do with the drawing in worsted. The drawing of worsted is performed by four or five sets of rollers operating upon the sliver in succession. Having got the condenser bobbins into the mule, and the slivers fixed to the spindles, let us witness the operation of drawing the woollen thread.

The moment the rollers of the mule commence to wind in sliver, the spindle carriage begins to recede from the rollers, bearing away the spindles, which are revolving and putting twine into the sliver preparatory to the commencement of the drawing. The rollers continue to wind in sliver until they have wound in about half the length, more or less, of the draw of the mule, and then they stop, the spindle carriage bearing the revolving spindles with the sliver attached still continuing steadily to recede from the rollers. The result, necessarily, of this operation is that the sliver becomes elongated or drawn out to twice its original length, and you will please note that the rollers have performed no part of the drawing, but have simply wound in the sliver, and then stood perfectly at rest until the spindles, with their compound motion of revolving and receding at one and the same time, have completed the drawing of the stretch of thread; shortly after which the mule stops, but immediately afterwards proceeds to wind the draft or stretch of spun yarn upon the spindle, in the form of a cop, by which movement the spindle carriage comes home again, close up to the rollers, ready to commence the same operation afresh by having another portion of sliver wound in and the carriage again repeating its receding motion with its load of revolving spindles.

Thus we clearly see that the mode of drawing peculiar to the woollen thread is compound, that is to say, that it is not and cannot be drawn by rollers, as in worsted, but only by the point of the spindle, which, at the same time that the fibres of the sliver are being drawn out into a thread, gives a certain portion of twine necessary to the construction of the thread, and to counteract the strong tendency of each individual fibre to shrink and curl up, and disengage itself from its fellows. The introduction of the twist in this peculiar way, while the draft is going on, is also necessary to the proper arrangement of the fibres in woollen fashion, so as to form a woollen thread.

Let us further observe that as soon as the drawing operation commences, which it does the moment the rollers cease to wind in sliver, there is a general movement amongst the fibres of the sliver, each fibre moving in the most precise order takes up its position in the forming thread, and the position allotted to each is, with the greatest mathematical exactness, according to its length. The twine and the draft, acting in concert, force the fibres to move according to their length; the longest are forced towards the centre of the thread, and the next fibres in the order of length follow, and so on in like manner until the operation is finished, and the stretch of thread is completed. The fibres are forced to take up these positions in consequence of the drawing of the woollen thread being a simultaneous compound process, the twine going into the sliver at the same time as the draft is going on, the two processes acting in concert simultaneously, thereby interlocking fastest, by the twine, the longest fibres, which, as the drawing proceeds, inevitably necessitates and forces the longest fibres, as they glide through each other's embrace, to take a position at the centre of the core of the thread, by reason of the greatest twitch or strain being thrown upon them through the twine going in simultaneously with the draft. The longest fibres have no alternative but to crush and elbow their way towards the centre of the forming thread, and so constitute the centre of its core. This comes about because they are more frequently entwined round each other by reason of their length, and are compelled to seek the centre in order to escape, as far as possible, the pressure of the twine brought to bear upon them during the progress of drawing. The next longest fibres follow close upon the leaders, impelled to crush and elbow their way also through the crowd by the same law of necessity, till they reach their position close around their predecessors near the centre; so the commotion goes on amid the crowd of fibres composing the sliver till each, by the pressure of the twist and the draft, has found its allotted position in the structure of the thread. The draft and the twine being at work simultaneously in constructing the woollen thread, the two have to be adjusted to each other's action in point of pressure with the greatest nicety. The form of the draft-scroll has to be varied in diameter at different points to suit the twine at different portions of the draft, otherwise the thread while forming would either lock fast by the twine overpowering the draft, or the draft would outrun the twine, and the thread become loose and break down; so that it is a point involving great nicety of judgment on the part of a skilful spinner to adjust the twine and draft to each other in order that neither the one nor the other may be allowed to take the lead or gain the ascendency, but that each shall work on a balance with the other. This balance has to be strictly and equitably maintained in all parts of the draft to produce a perfect woollen thread. The twine is a constant factor; its rate is uniform. The draft is an ever-varying factor. The latter has to be adjusted to the former in a ratio proportionate to the diminishing bulk and increasing attenuation of the forming thread and the varying spinning properties of the material in hand. There is nothing so complicated as the woollen draft in the whole range of our textile tissues. A young wool, for instance, is very different in its spinning properties from a wool clipped from old sheep. Some wools are dry and harsh, others soft and kind, all of which properties affect the woollen draft. Some of these properties affect the draft at the commencement and some towards the end. The skilful spinner dips the open fingers of the left hand into the drawing threads, and allows them to strain against the sensitive flesh of the insides of the fingers, and by this means the strain of the twine on the forming thread can be judged with the greatest nicety by the skilful workman as he sways his hand slightly to the right or to the left, so as to bear against the drawing threads as the draft proceeds. There is often a great difference between one parcel of work and another in its drawing capabilities, and the form of the draft-scroll has frequently to be altered so as to keep the draft on a balance with the twine. Some material "sets" or locks fast with the twine much sooner than others in the early portion of the draft, and has to be dealt with accordingly, while other material has other peculiarities that have to be met by varying the draft at different portions of the stretch. The dye has not unfrequently a very considerable effect on the spinning capabilities of the material.

We have said before that the twine is a constant quantity or factor, and in commencing a fresh lot of work the woollen spinner has first to try a few threads. He will usually put a single condenser bobbin into the mule, and try the draft, having first put on such change wheels as in his judgment will be likely to suit the draft of the work that he is about to commence spinning. If the twine overpowers the draft he will put on a larger change wheel and quicken the draft up to the pitch that his trial leads him to think is requisite, and then try again. In the second trial he finds that in the main he is getting near to what is wanted, but the beginning of the draft is a little too keen, and towards the end of the stretch or draw gets a little too lax. In the wood scroll of the hand mule the draft could be much more easily and more completely adjusted than in the self-actor. If the forepart of the draft was a little too keen, a strip of leather could be slipped into the cord groove of the scroll, and by increasing the diameter of the forepart of the scroll in this way the keenness of the draft could often be sufficiently modified, and at other times allowing the cord to work down a little nearer to the nose of the scroll, and putting a next larger sized change wheel on, would bring the draft more equal between the beginning and the end. It would be a great improvement in the self-actor draft-scroll if the edges of the cord groove at the beginning of the draft were a little deeper, and a number of half circle and a number of quarter circle segments of wrought-iron were bent to the curve of the scroll and fitted into the bottom of the groove, and fastened down with set screws with counter-sunk heads like an ordinary wood screw, which could be put in quickly or removed by a screw-driver. Such a contrivance as this would often be extremely useful to the spinner when he encounters work that will not lend itself to the form of the scroll. The draft-scroll is parallel for about half the stretch, and then gradually tapers down to about one-seventh the diameter that it is at the parallel portion. The segments just named should be about $\frac{3}{16}$ to $\frac{1}{4}$ inch in thickness, and would require the outer edges at the ends to be nicely rounded off, so as not to cut and damage the cord in working. It is also a great advantage to the spinner to have the scroll groove some 12 to 18 inches longer than the stretch of the mule, so that the cord can be worked down close to the nose of the scroll, or withdrawn from the nose, more or less, as the nature of the draft called for. Frequently the draft can be made equal in this way alone. A little extra length in the scroll to change about in is a great convenience to an able spinner, and in ordering new machines this should not be lost sight of.

While on the subject of drawing we may remark that in a bulk of wool there is always a number of bright slippery fibres that are troublesome to the spinner, and that prevent him setting the draft quite as keen as he would otherwise like; these troublesome fibres make known their presence by causing the thread to suddenly slip in the drawing, making a small or thin place, which is technically called a "twit." The "twit" is a thin, weak place in the thread; but no sooner does this slip take place in the drawing than, quick as lightning, the twine rushes into the weak place in the thread and stops all further drawing at that particular part in the forming thread until the remaining parts get reduced to an equal thickness with the "twit" or thin place; and not till then will the twine release its hold and withdraw its protection from the weak and enfeebled portion of the forming thread. This is a very striking and very interesting peculiarity of the woollen draft, and by observing and taking advantage of it the writer has produced some very pretty effects in the spinning and preparation of yarn for the fancy woollens, which are referred to on pages 347, etc. This action of the twine in rushing into and protecting the thinnest part of the thread during the compound draft, and, as a consequence, bringing greater pressure to bear on the more bulky or thicker portions, and compelling those thicker portions to reduce themselves down to the common level, gives the mule an immense advantage over every kind of frame as a finishing spinning machine, as it can smooth down the faults of roller-drawing both in worsted, cotton, and other textile threads. The twine holds and protects the weak portions of the thread, while it brings its strength to bear upon the more bulky and unruly portions that defy the action of the roller-drawing, and compels those more bulky portions to put themselves into better and neater order.

For the benefit of the student we may now sum up what we have seen and described as to the mode of the structure of the woollen thread. We have seen that the two blind forces, *the twine* and *the draft*, when directed to act simultaneously and in concert to attain a certain purpose, can sort out and classify the fibres according to length, and that instead of laying the fibres strictly parallel, as in worsted, it lays them lengthwise in a corkscrew form, and in graded order, so as to form the body of a thread that shall both be strong and elastic, and at the same time shall leave each end of each fibre standing out from the body or core of the thread forming the fringe that is the great necessity and distinguishing peculiarity of the woollen thread. We have seen that each fibre is made, in strict equity according to its length, by the strain and pressure of the twine, acting in concert with the draft, to take up the position to which it is entitled by its length. From the centre of the core of the thread to the surface there is this beautiful and exact gradation of the fibres or hairs of wool, each in the order of its length, until the shortest fibres become too short to be able to reach round the core, and consequently lose their grip of the revolving thread, and fall to the floor as *fly*, or what is technically called "fud." Remember also that the drawing of each separate length of sliver which has been wound in at each stretch of the mule has had to be completed into a length of finished thread or yarn before any additional sliver could be admitted. It is the admission of the sliver in this intermittent manner, and the completing of the drawing of the portion so admitted by the simultaneous drawing and twining of the spindles, that constitutes the product a woollen thread. A very erroneous notion is, that much drawing tends to making the woollen thread more like worsted; a more fatal error than this cannot be conceived in regard to the woollen thread. The following paragraph recently appeared in a journal of considerable standing: "*The fact must not be lost sight of that the woollen yarn may, and very often is, made to partake of the nature of the worsted yarn by drawing it out to its required fineness on the mule rather than condensing it to fine counts, which, of course, necessitates little drawing, and consequently little opportunity for the fibres to attain any great degree of parallelism.*" A statement more false or further from the truth and facts of the case could not be written. The woollen draft is much more complex than the author of the paragraph has any conception of ; indeed, he seems to have no idea of drafts, either worsted or woollen, or the principles that underlie each of them. Though it is drawing and twining that constitutes a worsted thread, and drawing and twining that constitutes a woollen thread, it does not follow from this that the two threads are alike because their constituting agents are individually the same. The two threads are diametrically opposite, though the constituting agencies employed in their structure are *individually* the same. All the differences between the two threads arise from in one case employing the agents separately and following each other, and in the other case employing them jointly and in concert at one and the same time and place.

My experience of the practical working of the woollen mule for a period of fifty years, and close study of the principles of the woollen and worsted drafts in the meantime, have led me to a very different conclusion from that embodied in the above extract, as no one could make such a statement expressing his view of the matter except through utter want of all elementary knowledge of the principles that underlie each of the drafts of worsted and woollen, the worsted draft producing a thread with the fibres as parallel as it is possible to lay them, the woollen draft producing a thread with the fibres as far out of the parallel as they can possibly be arranged consistent with the production of a sound textile thread that shall be able to stand the strain of the loom, and capable of a proper felt and finish, and that shall give satisfaction in the wear and tear of the after use of the fabric.

The drawing of the woollen mule does the very

opposite to what the above extract says it does. The drawing of the woollen mule takes every fibre out of the parallel and puts it into the woollen form, and the more woollen-drawing we bestow upon it the more woollen it becomes.

So far from the woollen yarn being made to partake of the nature of worsted yarn by drawing it out to its required degree of fineness on the mule, the absolute contrary is the fact. Drawing on the woollen mule does not cause the yarn to partake of the nature of worsted, but actually prevents the fibres taking up the worsted form of arrangement; and not only that, but where the worsted form of arrangement already exists, the woollen-drawing immediately breaks it up and transposes the arrangement of the fibres into the woollen form. I have taken bobbins of worsted roving and put them into the woollen mule, and, by three short, gentle drawings, have taken the wool fibres out of the worsted form and transposed them into the woollen form, and produced a fairly good woollen thread by subjecting the worsted roving to woollen-drawing; so that woollen-drawing not only prevents all tendency in the wool fibres to assume the worsted form by long drawing, but actually breaks up any parallelism or tendency towards that form wherever it comes into contact with it. Simple drawing alone would, and always does, lay the fibres parallel, but when the drawing is not simple, but is a complex and compound drawing like the woollen, consisting of drawing and twine in simultaneous action, the action then becomes a double or compound one, and the effect of this double action upon the arrangement of the fibres is quite different from what it would be if the two movements constituting this double action followed each other separately or consecutively as simple movements, and not acting in concert at one and the same time. This double action of two forces, made to act in concert, though opposed to each other in nature and individual tendency, produces effects quite the opposite of what would take place if each were acting separately without the restraining action of the other.

The woollen thread that we have attempted to outline cannot be produced by the amount of drawing that can be obtained by passing the material only once or twice through the mule. A first-class woollen thread requires to pass through three or more operations of drawing to complete its construction aright, and we at this period ought to have advanced the drawing from the ten units of the old billy days up to at least fourteen or fifteen units, by putting it through the "slubbing," "roving," and "finished spinning" processes, all of which are really spinning in its different stages of progress.

The condenser in its very nature is alien to the woollen manufacture, and, used in its present-day form, it is literally a curse, as it deprives the thread of its drawing, which is the very foundation of its existence and the basis of its structure. There would be a much better chance of producing something nearer a woollen thread if, instead of taking 120 threads from a 48-inch condenser, people would be content to take off only 20 threads, and give the other 100 threads back again to the drawing from which they have been so ruthlessly stolen. We should then see some signs of returning sanity on this very important matter of drawing in the spinning. This may appear to be strong language, but it is not any stronger than the situation calls for, as the woollen thread is built upon drawing, and upon drawing only, so that anything that deprives it of any portion of its due amount

of drawing does it an irreparable injury. If the
worsted people find it necessary, and to their advantage, to bestow all the series of drawings that we
have alluded to in order to perfect their thread, will
it never dawn upon the woollen people that there is,
perhaps, a principle of structure in the woollen that
may require a series of drawings also, in order to properly
construct a woollen thread? The condenser, by depriving
the thread of nineteen-twentieths of its former drawing,
has been the very *death-blow* to the best interests of the
woollen trade, as it renders the production of all cloths
that require a fairly well constructed woollen yarn impossible. For instance, the decline of the superfine black
cloth coating dates from this period; the same can be said
of the doeskin trade. So long as the billy yarn, with
its twenty times as much drawing as the condenser
yarn, was used, such cloths as the superfine west of
England coating could be made, but immediately the
condenser yarn was used the cloth was found to be
completely changed in character. The production of this
cloth is a severe and very sensitive test, and any radical
change, such as that made by the condenser, is at once
indicated in the cloth. The tissue spun direct at one
drawing from the condenser was not a fully developed
woollen thread, therefore when put into a superfine west
cloth the fabric was not of the normal strength; the cloth
was not sufficiently full in cover for the finisher; it was
papery to handle, and destitute of the usual suppleness
of the best west cloth. Every attempt was made to
remedy these defects by the finishing, but in vain. To
overcome the want of cover or balk, more raising was
resorted to by the finisher to get up the usual pile and
lustre of face, more boiling was tried, but it only further

tendered an already too tender cloth, and every effort failed to mellow it, and the face had a gray, washed appearance.

The final outcome was that the cloth was tendered by excessive raising to get up the pile from a ground incapable of yielding it to form the " face," and the additional roller boiling only added to the mischief ; the proper fabric was not there to finish.   The extra efforts of the finisher only tended to make the cloth rotten, gray on the face, papery, and stiff.

Give back to the woollen thread its drawing in full measure without stint, and then, with a thread " slubbed," " roved," and " spun," we may resume the production of the west cloth with success, or any other cloth that our taste or fancy may lead us to desire ; such a thread would bear a thousand applications that we at present have no conception of, and the trade would be constantly finding fresh outlets by its improved and further-developed thread, just as the worsted by its improved thread has found fields of use never before dreamed of.   With a fully-developed woollen thread the manufacture might hold up its head once more and go forward, as in addition to a super-faced cloth, should public taste call for it, an immeasurably superior class of tweeds could be produced to anything at present in the market.   Our present success is quite equal to our deserts, as we bring neither labour nor brains into the field in comparison to our friends in the worsted industry.   We slip the material through the condenser and once through the mule, in a kind of slovenly manner, and consider we have done everything that any reasonable being could possibly demand of us.   We have thrown aside the woollen thread of our fathers as a thing of naught, in order to follow the "will-o'-the-wisp" delusion of the Yankee "condenser," and in doing so we have our reward in getting a thread with only one-twentieth of the drawing in it; and in thus senselessly following the delusion, we have not shown even common English shrewdness, or any regard whatever for the best interests of the industry we are engaged in, to say nothing of the want of science or principle involved in such a step. To stand still is fatal in these stirring modern times, but to fall back is doubly lamentable. We have lost 95 per cent of our drawing by adopting the condenser. If we had only had the sense to retain the thread of our fathers' day, and, instead of falling back, had kept right on abreast of the times, by now we might have had a woollen thread possessing something like twenty units of drawing in its structure, and with such a thread at our command the possibilities of the woollen industry would have been great, and we could have rolled back the "bore" of any opposing tide, and might now have been in possession of a prosperous manufacture.

The men that introduced the condenser did not understand the structure of the woollen thread nor anything of the disastrous consequences that would follow their dispensing with the drawing at the billy or its equivalent. No one, understanding clearly the nature and use of drawing in the structure of textile threads, could have been so infatuated as to cut off so very large a portion of the drawing, amounting on an average to nineteen-twentieths, or 95 per cent of the whole of the drawing in use at the time. The sliver at the time of the billy was taken from off a card-breadth of 5 inches, sometimes more and sometimes a little less, and then reduced by drawing, as in other textile industries; but in the modern condenser, the card-breadth from which the sliver is taken is less than $\frac{3}{8}$ of an inch, even on a Bolette condenser. The Martin lap condenser goes farther than the Bolette, and takes 120 from $45\frac{1}{2}$ inches wide. This being the case, there is no room to get in any drawing worth mentioning, as the sliver, before it touches the spindles, is in most instances within half a unit of the counts required, so that there is nothing more than a mere shadow of drawing in the structure of such a tissue, and it is a base application of the English language to call such a shadow a thread. A proper and substantial thread can only be built up by its requisite amount of drawing from the thick sliver, and the mode of drawing to be adopted is determined by the character of the thread required, whether worsted or woollen. *All our textile threads* have to be taken in the *thick sliver* from the *carding,* *combing, heckling,* or other preparing machines, and then reduced by drawing. In the reduction by drawing roller-drawing is resorted to if the material is for linen, cotton, or worsted; but if the material is for woollen, then spindle-drawing is resorted to. In each instance the specific mode of drawing adopted is that which accords with the nature of the thread required. In every instance reduction by drawing from a thick sliver is resorted to in order to afford the opportunity of arranging the fibres in such a manner, parallel or otherwise, as the nature of the parti- cular thread that is in hand requires; and without this reduction of the thick sliver and arrangement of the fibres by drawing there can be no proper textile thread; therefore to reduce the sliver by the condenser instead of by drawing is *contrary to all common sense, contrary to all science,* and *contrary to all and every principle of textile thread structure.* We have already reminded the student that since the introduction of the Arkwright frame, a little more than a hundred years ago, we have had two different principles of spinning at work, but whichever of the two principles of spinning we turn our attention to, we find that the forming of a true and pure thread always begins

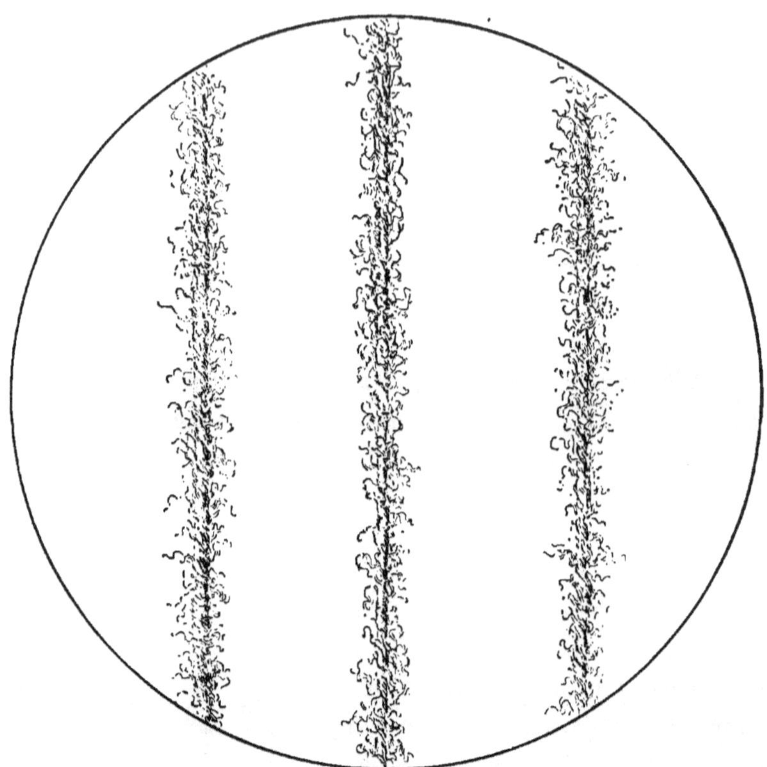

FIG. 57.—A WOOLLEN THREAD, FROM THREE FULL DRAWINGS.

in a thick sliver, and *the fibres are arranged as the reduction by drawing proceeds, each thread after its kind.*

This beginning of the production of a perfect woollen thread from a thick sliver and reducing it by drawing is an absolute necessity, and is a mode of procedure analogous to that followed out in every other kind of textile thread.

It is found to be necessary to begin the production of the worsted thread, as well as the cotton, flax, hemp, spun silk, etc., with a thick sliver, and reduce it by drawing by rollers, in order to get the fibres in each case laid parallel;

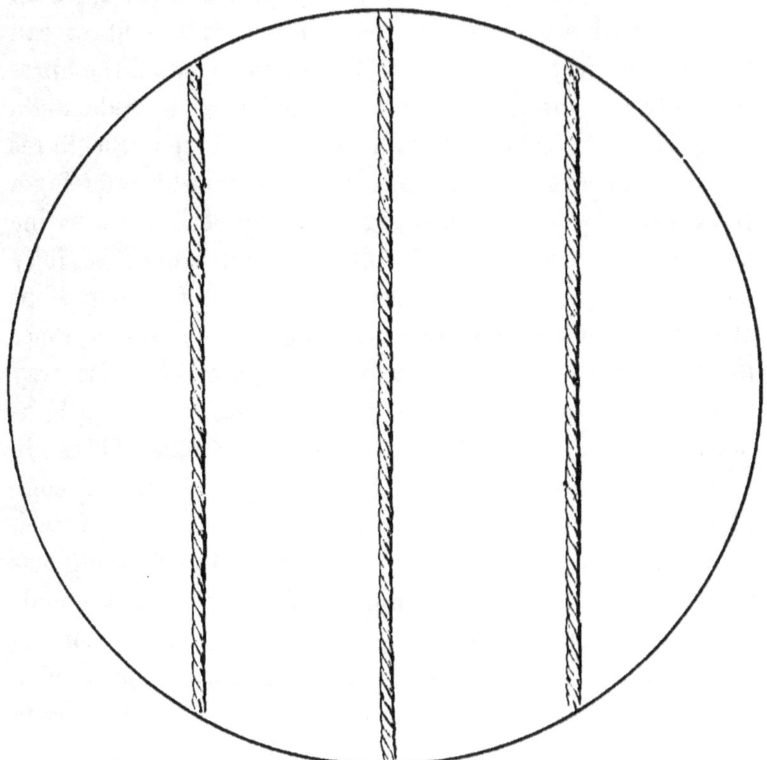

FIG. 58.—A WORSTED THREAD, TWO FOLD 48s.

and only by following out a course of drawing of an opposite character, that is, drawing by the spindle, instead of by rollers, can we produce the woollen thread. Only by spindle-drawing from a thick sliver can we produce a thread that is not bare, like the worsted or cotton threads, but that shall show the ends of the fibres standing out from the core or main body of the thread in a chenille-like fringe. It is this fringe, composed of the ends of the fibres, that makes the woollen thread so valuable for all branches of the woollen manufacture, whether plain or fancy ; and it is only by a series of spindle-drawings, each drawing with a reversion of the twine, that the fibres can be arranged in this form. Just as in the worsted the fibres can only be caused to fall into parallel form by reducing a thick sliver by roller-drawing, so in the woollen the fibres can only be caused to fall into woollen form by reducing a thick sliver by spindle-drawing, and by reversing the twine at each operation. The woollen thread cannot be fully developed by either one or two operations of drawing, but at each operation a portion of the ends of the fibres comes to the surface as the body of each fibre finds its way towards the centre of the core, according to its length, as before intimated. The arrangement of the fibres in longitudinal corkscrew form is nearer and nearer complete at each operation of drawing, and the ends of each fibre stand off in fringe-like form when the drawing has absorbed the body of the fibre in the twine, till the ends have become short and unable to reach round the core of the forming thread. The complex nature of the woollen draft when in operation absorbs the body of the fibres to form a core while it at the same time liberates the ends. When the carded wool is made into thick sliver there is no arrangement of the fibres, and the ends are mostly embedded in the bulk, but as the woollen draft proceeds, the draft and the twine, acting in concert, force the body of the fibre into forming the core, and the ends into forming the outside fringe. This curious and wonderful arrangement of the fibres in a woollen thread is brought about by the simultaneous action of the draft and

twine and the balancing of each of these forces against
the other, and as all through the draft there is every
inch of the way a constant tendency in the draft to
get off the balance, it requires the greatest nicety and
judgment in its adjustment, and the utmost constancy
of mechanical action, to carry it on when once adjusted,
and the eye and hand of the skilful workman are
constantly needed, as the stretching of a cord or any
other trivial thing may at any moment throw the draft
off its balance, and call for readjustment. It is this
mechanical constancy in the draft of the mule that makes
it such a superior spinner to the old hand jenny, not that
its draft is any different in principle, but that it is purely
mechanical in the mule, and therefore uniformly constant,
which could not be said of the old hand jenny, even in the
hands of the most skilful workman. The great uniformity
in the action of the modern mule enables spinning to be
carried to a very high degree of perfection. The draft
has, as we have said, been purely mechanical since the
early years of the mule, but the superior build of the
modern mule enables the draft to be carried out with much
greater nicety and unvarying exactness than was formerly
possible. The spindle carriages are much better built in
every way, and far more exactly laid down on the slips
than used to be the case, and as a consequence they
move with much more steadiness than formerly. The
slips used to be laid with great irregularity, but now a
spirit level is used and one uniform fall allowed to every
slip, so that there is one uniform strain throughout the
length of the carriage, which was far from the case during
the earlier years of the mule, besides which the head-stocks
are much stronger and far more truthfully fitted, even in
the later makes of the hand mule. Every one of these items

contributes a part to the general result, and enables the modern mule to carry out the woollen draft with a degree of exactness that was utterly unattainable in the mule of forty or fifty years ago, with its loose build of carriage and untruthfulness in the fitting up and setting for working, all of which contributed to make the carriage move in and out with a very unsteady, "jerky" action that much interfered with the conducting and adjusting of the draft to that steady uniformity and exactness that is so desirable and necessary.

It seems never to occur to the manufacturer that in following out the condenser craze to such an extreme as is exemplified in condensing the twenty-four skeins yarn to a sixteen skeins sliver in the first instance, and allowing only a half-unit of drawing in the spinning of it to twenty-four skeins, there is little more than a shadow of textile structure in it, and that the thread, or rather the apology for a thread, is only held together temporarily by the gird of the twist, like a farmer's hay rope. Should the cloth, with so little spinning-drawing in its thread, be intended for a full-milled faced finish, before the fabric is halfway through the fulling or milling process the little thread texture that it possessed is practically all dissipated, and the fabric to all intents and purposes becomes nothing but "a felt," in comparison to what a real textured cloth ought to be, its texture all or nearly all gone, and it is only held together by the felting that it has received in the fulling mill, instead of there being in the foundations of its structure a thread texture to give it strength, suppleness, and every other property that a rightly-constructed cloth made from a rightly-constructed yarn possesses. Need we wonder that a cloth from a yarn with only an apology of structure in it is tender when finished? It would be marvellous indeed if a fabric so constructed were not tender or destitute of strength. There can be no strength without a suitable foundation texture, and there can be no texture that will resist the disintegrating power of the fulling mill unless it is composed of a thread that is solidly and durably compacted together by having had bestowed upon it the full complement of spinning-drawing. Such a thread, and such only, can constitute a proper foundation for a durable and in any way satisfactory texture in a woollen fabric. With a foundation texture so composed we shall not only ensure strength and durability, but we shall ensure the attainment of cover or balk by milling, along with suppleness, mellowness, fulness in the hand, as well as freedom from that washy grayness that ever accompanies a fabric made from a crude and partially-drawn yarn. There are no effects without causes, and if we want to see certain effects in cloth structure we shall have to take the trouble and pains to put foundations into the cloth adequate to sustain the production of the effects we desire. We have no right to expect certain effects in cloth manufacture, or in any other manufacture, unless we have the skill, judgment, and technical and scientific knowledge requisite to organise, set in motion, and direct the necessary causes that will in the ordinary course produce the effects desired. If we put no more thread structure into the texture of our cloth than the milling process is capable of dissipating, we have no right to complain if the cloth turns out too tender and unsatisfactory when it is finished; it is simply our own fault in not having a thoroughly grounded knowledge of the manufacture that we are engaged in; we have nothing to blame but our own want of knowledge. The reaping has been according to the sowing; what more do we want, or what more have we a right to expect? We have no right to expect more structural effects to arise out of the cloth than we have laid in foundations for. To have texture and strength in the cloth when finished we must have structure put into the yarn that is to compose the texture at the proper time while the yarn is being spun. There is only one way of obtaining strength and suppleness along with elasticity in a woollen cloth, and that is in having a yarn with the full complement of spinning-drawing in its structure. Without this there is no cohesion and unity of action in the fibres of the thread to sustain any amount of strain, but with the fibres arranged as they are in a properly-constructed thread there is a marvellous amount of elastic resistance to strain, as the fibres band more closely together, right up to the point of breakage, in proportion to the strain put upon them. Our super west cloths are all tender or wanting in strength when finished. The best of them are tender in comparison to what they used to be before the introduction of the condenser; the extreme length to which the sliver is now being condensed, and the consequent displacement of spinning structure of the thread. If we put no structure, or a very inadequate structure, into the thread as it goes through the spinning, we are very unreasonable in looking for effects that are only obtainable through full structure in the thread. Effects can only spring from adequate causes, and if we neglect, or have not the technical knowledge and skill, to set those causes to work, we shall not find the effects that we want to find at the close or finish of our manufacture. It is vain to expect any such thing, as we have failed to do that which alone could produce, or bring about, the effects wanted. The very little structure of thread that can be obtained by one very partial spinning-drawing after the condenser is soon scattered to the winds in the milling process, and the fabric becomes a solid felt, all texture gone, or nearly so, and hence arises the stiffness, the paperiness, and the tenderness of our present-day woollen fabrics; whereas a properly-constructed thread makes a cloth that never loses its texture in the milling process, however hard it is milled, but keeps that texture intact; and milling only compacts it, and the balk forms on the surface of the cloth, both back and face. It does not balk through, but its interior is a close texture, and from this texture of the fabric we derive the suppleness, strength, mellowness, the fulness and evenness of balk upon the surface, and that closeness of face when finished which the partially-formed thread of the condenser can never produce. Without an abiding texture we can never obtain suppleness or strength, and without a fully-drawn, fully-covered, and in every way well-formed thread we can never obtain a satisfactory texture of fabric that will withstand the milling and finishing operations unimpaired. A texture from such a thread will resist the balk going through the cloth, and confine it to its proper place on the surface, and thereby obtain the suppleness, mellowness, strength, and beauty of a true cloth.

The above remarks may at first sight appear to apply more particularly to full-milled faced cloths, but they are equally applicable to our fancy woollen cloths, as their great defect at the present day arises from the want of cover in the yarns used in their production, and this in turn arises from want of spinning-drawing. This want causes a hard, bony handle in the goods, as well as a want of elasticity, strength, and suppleness, and prevents the bringing out of the colours effectively. All these defects arise from one and the same cause, whether met with in the finest west of England super or in an ordinary fancy woollen, and that is the want of a properly full-drawn yarn. It is a radical or cardinal defect, and never fails to make its presence known, whether in plain super goods or in fancy woollens. The displacement of the spinning-drawing by the condenser, and striking out the billy-drawing altogether, was a fatal error, and stands in the way of all excellence in our fancy woollens, as well as in our super plain cloths. Neither style of pattern nor quality of material can make up for the defects that arise from this cause. Indeed, a full-drawn thread is of such importance in every branch of the woollen manufacture that it is a primary condition of all excellence. Without the full complement of spinning-drawing we can neither get cover on the thread, nor, as a consequence, strength, elasticity, suppleness, pile, or velvety handle, full effect of colours, nor full effect of pattern. There is always a crude rawness about a cloth produced from a partially-drawn thread, both to the eye and to the hand; it is bony to the touch, and naked and repulsive to the eye.

Nothing will lift the woollen manufacture out of the mire but excellency of production; nothing but this will enable us to retain our hold of the neutral markets of the world, and nothing but excellency of production will enable us to break down the hostile tariffs of the world, and to open out *new markets* for the products of our manufactories. We ought to depend upon nothing less than the superiority of our productions and their price as the foundation of our future progress, and this state of things can only be brought about by the attainment and active exercise of the very best scientific, technical, and practical knowledge, working diligently side by side, ever bearing in mind that anything

that is scientifically wrong will never work out to be
practically right. The woollen manufacture is at a great
disadvantage at the present time, and we repeat, *and challenge proof to the contrary, that the woollen thread was never as
badly and as imperfectly formed as in the closing years of this
nineteenth century.* There never was in the history of the
woollen manufacture so great a need of the best scientific
and technical knowledge to be brought to bear upon it in
its earlier processes as there is now, beginning with the
fleece, and following it step by step right up to the
loom. Every process is in the greatest need of a thorough
overhaul, and of the best scientific and technical help.
There never was so great a need to revert to foundation
principles in order to weed out the many errors that
have crept in during the last half-century, and to restore
the manufacture to a healthy condition by teaching the
elements to the classes in our technical schools and
colleges, and to place it on a sound scientific footing to
enable it to face the future. The great division of labour
during recent years seems to have operated adversely in
isolating one process from another, and in obscuring the
connection that exists between the respective processes, and
even further, in losing sight of the principles on which the
respective operations rest. While these isolations have
been taking place through the division of labour in our
mills, alterations and changes in machinery have followed
each other in such quick succession that in many instances
the rage for change from old machinery to new has been so
rampant that in most cases the alteration has been contrary
to the scientific principle on which the process was founded.
The result of the change has been a very marked deterioration, as is strikingly exemplified in the case of the
introduction of the condenser, where the spinning-drawing

is thrown away at a point and at a period in the manufacture when it ought to have been increased. There is no remedy for such a state of things when they occur, and no prevention of recurrence, save in teaching the elementary scientific principles on which each process rests to the classes in our technical schools and colleges; then we should be spared these lapses or driftings away from first principles from which the manufacture is suffering so severely at the present time. We cannot cast away such a large proportion as 95 per cent of the spinning of the woollen thread, so as in a sense to practically destroy its manufacturing value, without bringing upon ourselves the consequences sure to follow the use of so seriously deteriorated a production.

If, at the time we refer to, such elementary, scientific, and technical knowledge had been widely diffused among all classes engaged in the woollen industry, we should not have had the condenser foisted upon us with all its train of dire evils. We need to have this elementary, scientific knowledge of first principles taught and widely diffused, not only as a means of progress, but, as we see to our cost, *as a preventative of retrogression.* A general diffusion of scientific knowledge respecting the structure of the woollen thread would have prevented every such retrograde movement as that which was ushered in by the condenser, as it would have been clearly seen that such a move would end disastrously to the best interests of the woollen manufacture, and it would have been also clearly seen that to deprive the thread of its spindle-drawing was the most fatal step that could possibly be taken, seeing that it is this kind of drawing that constitutes its existence as a woollen thread, just as drawing of a different kind constitutes the worsted thread. A better diffusion of scientific and technical knowledge would have shown that all the openings for improvement lay in the direction of *increased* drawing instead of in the direction of diminishing it, and when the nature and structure of the woollen thread are better understood and its value properly appreciated—when its nature and structure become a subject of class study in our technical schools, then the three drawings of *slubbing*, *roving*, and *finished spinning* will be deemed insufficient, and a series of four or five drawings will be considered not too much. It unfortunately happens that our technical schools have done comparatively nothing for the early processes of the woollen manufacture before the weaving stage is reached : a little has been done for the dyeing, but even there much yet remains to be done. The other early processes have received scant attention hitherto, but have been open to all sorts of senseless inroads by people without knowledge of the scientific principles that are involved in the manufacture. While passing this point we are quite willing to admit that the best technical education in the world cannot supply the place of an earnest painstaking desire on the part of the student to understand. Technical education can only help at the utmost : it cannot lead, but it can help those who ardently desire to lead, and to lead intelligently, and avail themselves of the best modern mechanical and other appliances pertaining to the manufacture.

If some intelligent and enterprising manufacturer would only take up the woollen thread at the point where it stood fifty years ago, and add 50 per cent more drawing to it, we should then get a modern woollen thread of some considerable value in manufacturing capacity, and this could easily be done by taking the carding or sliver from off a card-breadth of 7 inches, instead of the old 5-inch breadth

of fifty years ago, and piece the cardings by the piecing
machine as before. Besides this, much might be said in
favour of adopting the transverse mode of doffing the
wool from the carding machine, as we then get the
wool in its best form for most quickly producing a
thoroughly perfect woollen thread. When we take the
wool off the cards by the condenser we take it in its worst
form for most quickly producing a good woollen thread,
that is, with the least possible amount of drawing, as by
taking it off with the condenser we take it off lengthwise of the fibre, with a certain amount of parallelism
in it, and as a consequence part of the first drawing is in a
manner wasted in taking the wool out of this partially
parallel form before the work of putting the wool fibres
into the woollen form can be really begun. On the
other hand, by doffing the wool transversely from the
carding machine we get it in the form most favourable for
commencing at once, as soon as the drawing begins, that
arrangement of the fibres which constitutes the woollen
thread. By taking the wool fibres off transversely from
the carding machine we get them without any parallelism,
and none of the drawing is wasted in taking them out
of the parallel form before the woollen structure can be
commenced, but the drawing goes at once to the forming
of the fibres into the arrangement that constitutes a
woollen thread. Doffing the wool transversely from the
carding machine is in every way most in harmony with
the principle of the woollen thread structure, and very
much might be said in favour of that mode of procedure, as it is a material point gained if none of the
drawing has to be expended in breaking up the previous
arrangement of the fibres, but the whole utilised in commencing and finishing at once the woollen arrangement

of the fibres, avoiding all loss or waste of drawing before the woollen formation can begin.

This idea or plan of doffing the wool transversely can be very economically carried out, and will give the manufacture a thick sliver in a convenient form from which to commence drawing earlier in the process, and will also give a sliver with the fibres in the best form for conducting the drawing to the best possible issue and the production of the most perfect woollen thread. We need not revert to the billy for conducting the slubbing process, as the slubbing is only a thick spinning or first spinning, and can be better and much more economically done on the mule. The billy disappeared from the cotton manufacture, for which it was invented, long before it was laid aside in the woollen, yet the slubbing still goes on in the cotton by other machines, and the same course can be taken in the woollen. The condenser bobbins could be taken to a piecing machine, which would piece the thick transverse rolls or cardings that we have just been mentioning, and the full bobbins could be taken to the mule and there slubbed after making slight alterations in the mule, such as making the little top rollers over the fluted ones double the weight of those used for finished spinning. The surface winding drum in the piecing machine would require to be adjusted to the width of the bobbins, or *vice versâ*, as most convenient. In the mule, if under $2\frac{1}{2}$ in pitch, every alternate spindle could be used for a time while any one was making a trial of a few months, and the spindles and other matters could be definitely arranged afterwards. One mule would be able to follow two, or in some cases three sets of carding machines, and any extra cost of production would be small indeed in comparison with the advantage that would be derived. The slubbing cops could be taken and *roved* on any ordinary mule provided with back creels. A forward movement of this kind would result in the production of a thread with thirty times as much drawing in it as the present condenser thread, and with thirty times its manufacturing capability ; and if the woollen manufacture were to take such a step as this it would not be doing any more than the cotton and the worsted industries have each done during the last fifty years. Look again at the instance given, some pages back, of the amount of drawing bestowed upon a good quality of 48s. worsted in Bradford, and corresponding figures showing the great amount of drawing in the cotton could be cited also. Without a large amount of drawing there can be no thread produced of any high manufacturing value. Then why should we have ruthlessly and ignorantly thrown away nearly the whole of our drawing instead of increasing it ? Why should we still hesitate, or why begrudge the production of a proper woollen thread when two such large industries as those of the cotton and the worsted can be cited in favour of such a course of forward movement ? To stand still during fifty years is bad enough, but to fall back as the woollen industry has done is worse. If our worsted friends, in face of the keenest competition, can have screwed their courage up to bestow ten thousand millions of crossings and re-drawings upon the worsted thread, can we in the woollen, *with any show of reason*, refuse to bestow the three drawings of *slubbing*, *roving*, and *spinning* upon our woollen thread ? When the structure of the woollen thread is better understood, and its value properly estimated, we shall not hesitate to bestow more than three drawings upon it. Drawing is the only condition on which such a thing can be obtained, and if we think otherwise we lack a knowledge of the most rudimentary and simple elements of textile thread structure. We cannot get a thread worth the name with the amount of drawing that can be obtained after the condenser, and it is very unwise to limit the drawing so as to get only a partially formed thread, which is of little or no manufacturing value. The wisest policy is to bestow as much drawing as is requisite to constitute a well-formed and fully-developed thread, perfect in its kind, as the worsted people have done, and this cannot be done at the lowest computation with less than three separate and distinct full drawings—slubbing, roving, and spinning. The cost of *slubbing* and *roving* is small in amount, but when compared with the advantages it yields it is not worth taking into account in medium and best-class goods. The highest known standard of perfection that is attainable should be the first consideration in all qualities of the manufacture, but especially should this be our object when we reach medium and high-class goods. Unless the cost is excessive, and is such as will exclude the manufacture from the general market, which is not the case here, it ought in the first instance to be set aside until the article is produced in the best known manner that is practicable; then let cost be looked at, and strive to produce the article at the lowest possible price. Produce the article first in the best known manner, and then produce it as economically as possible, and in the majority of cases a market for it can be found. Slubbing and roving, rightly conducted, will more than repay cost.

All the three stages of spinning just named can be done on the mule, and all will require to be spindle-drawings. Without a good amount of drawing there can be no production of a textile thread, as is shown by the worsted and cotton, except in such an instance as that of net silk, which is a naturally formed thread, and only needs doubling. All artificially formed threads can only be constructed by means of a considerable amount of drawing, as is illustrated in the production of worsted, cotton, flax, spun silk, alpaca, mohair, etc. The amount of drawing suggested as being comprised in the three operations of woollen spinning, drawing, slubbing, roving, and finished spinning, is small indeed in comparison to the amount of drawing that has been cited as being used in worsted at the present time. The woollen industry must, indeed, arouse itself, and not boggle at such a trifle as the amount of additional drawing that may be required to advance a stage forward from the position occupied fifty years ago. Before much good can be done we must make up our minds to take the thread structure at the point where the old billy left it forty to fifty years ago, and add at least something like 50 per cent to the drawing of that date. The roll of carded wool, or the carding, was taken from a 5-inch card-breadth in many instances, and to obtain the additional drawing named we should require to take the carding from a 7 or $7\frac{1}{2}$-inch card-breadth at least, and thereby commence the drawing earlier, and at a thicker stage, so as to allow scope during the drawing for the fibres to get more perfectly arranged in the woollen form before the finished spinning stage was completed. A much superior thread could in this wise be produced than any we have yet had in the woollen manufacture. The worsted thread of the present day absorbs infinitely more drawing than anything that has ever been asked for or seen in the woollen, and to ask that the woollen thread may be brought nearer abreast of the worsted in perfection of structure surely cannot be thought for a moment unreasonable or untimely, seeing in how many instances the worsted has displaced the woollen during recent years. Certainly more drawing will have to be bestowed upon the structure of the woollen thread before the industry can again take up its rightful position among the great industries of the empire. We shall not only have to give back the drawing of which we have so unwisely and so unscientifically deprived the woollen thread by the condenser during the past fifty years, but we shall have to take a step farther forward, in addition to bringing the thread simply back to the position that it held fifty years ago. The position that our forefathers attained to (and which we have lost) ought not to satisfy us; we ought to make an advance of our own, and leave our own "footprints" for the better "on the sands of time." Each generation has its own obligations and duties to perform in respect to making advances and improvements upon the previous one, but the present generation has a double duty to perform; it has not only a generation's duty to perform in respect to obligations to contribute its quota of improvement for the benefit of the race, but it has the additional duty, as far as the woollen industry is concerned, of recovering that which it has, for want of better technical knowledge and more thoughtful attention, so unfortunately lost as regards the structure of the woollen thread. Let us immediately retake the position that our forefathers had attained to in the woollen manufacture, and then make our own advances and improvements thereon. Until this is done let there be no boasting of our modern or rather present-day improvements in woollen manufacturing, but let us conduct ourselves with becoming humility. Our present woollen thread is nothing more than a pretence, a cheat, a swindle, a fraud, palmed off as a woollen thread, when compared with the properly constructed twin-wool thread of the worsted.   The worsted people expend a great amount of drawing upon their thread, and get a valuable product in return, and we can get a valuable product also in woollen by an adequate expenditure of drawing, not otherwise.

In the light of the worsted, our present-day condenser thread has literally no construction in it.   It is merely a sliver girded together by twine, and of no manufacturing use or value, in comparison to that of a fully-developed and properly - constructed woollen thread.   The manufacturing difference and capabilities are great between a properly constructed thread and a tissue that is nothing more than an outline form or shadow, like our present undrawn, unformed condenser product.

The one is so superior to the other that they cannot be compared.   Things must have some degree of resemblance to each other before a comparison can be instituted.

That which is so incomparably superior cannot be brought into comparison with such a wretched apology for a woollen thread as the condenser product.

The fatal thing in the adoption of the condenser was that it deprived the thread of its drawing—deprived it of that which constituted it a thread; therefore the objection applies to every kind and form of condenser whatsoever, without exception.   The principle of the condenser is in the highest degree pernicious and subversive of the best interests of the woollen industry; the more perfect the kind of condenser is, as a condenser, the more pernicious is its effects.   The reader is asked to weigh carefully and consider calmly the reasons given for such statements on the condenser as are made here and elsewhere in this text-book.   The statements are not made without sufficient reason, and are the result of long consideration and patient study of the condenser and its effects on the woollen thread structure. The want of cover in the condenser thread, in consequence of its being only so very partially formed, having only one-twentieth of the drawing in it that its predecessor, the old billy thread, possessed, has led to some amusing instances of ingenuity, or what would have been amusing were it not that they have a very painful side, in order to overcome the want of cover in the production of certain kinds of cloths. The utter lack of sufficient cover in the condenser thread through want of drawing has led in several instances to the use of *noils only* in the making of certain cloths. It may be said here, in passing, that noils are the shortest and most curly parts of the wool that are combed out of the worsted during its manufacture. The noils are combed out of the worsted in order that the thread may be clear and bare. The fine noils have had to be resorted to in the making of the cloths in question because they are the most curly material obtainable in wool, and have therefore been employed in the production of these cloths in order to get the cover required in an indirect way. Were the woollen thread of the present day constructed on a rational and scientific principle, like that of our forefathers, and that on which the worsted of the present day is constructed, we should have no need of such instances of ingenuity in overcoming the difficulties of the situation.

The two wool threads, woollen and worsted, when properly constructed, are both of them highly elastic; this arises in each instance from the perfect arrangement, each after its kind, of the fibres in each of the threads.

Let an old and skilful woollen spinner take a length of the modern condenser thread, say a foot in length, into his hands, and stretch the thread to the point of breakage, and he cannot fail to notice the starchy and unnatural stiffness with which it breaks. There is none of the elasticity natural to a truly constituted woollen thread, and this arises from the want of structural arrangement of the fibres, as the thread is only very partially formed. The breakage is sudden, harsh, and stiff, whereas the breakage of a true thread is slow, elastic, and kindly. And as is the breakage, so is the thread itself; when put into cloth it is papery, harsh, and non-covering, stiff, and unkindly in the handle. Instead of the miserable make-shift of having to resort solely to noils in order to get the required cover, in a thread constituted by the requisite amount of drawing a cover equally good, *or better*, could easily be obtained by taking 20 lbs. noils to 10 lbs. clean wool, and in addition to the cover of the balk that would be obtained we should have the *elastic toughness, the strength,* and *suppleness so essential to a true cloth.* It is the very nature, tendency, object, aim, and end of the woollen thread to produce cover, just as it is the nature, tendency, object, aim, and end of the worsted, by an opposite mode of drawing, *not* to produce cover, but bareness of surface. The more woollen-drawing you put into a woollen thread, the more woollen it becomes, and the farther it is removed from worsted, just as, on the other hand, the worsted people find out that the more worsted-drawing they put into their worsted thread, the more perfectly worsted-like it becomes; and they have had both the good sense and the scientific knowledge to follow up the principle into thousands of millions of crossings and re-drawings. Both classes of textile threads, roller-drawn and spindle-drawn, are based upon or are dependent upon drawing for their structure, and only by drawing can either the one class or the

other be produced. We have seen to how high a degree
the drawing is carried in the production of a good quality
of 48s. worsted, and we shall have to see the woollen-
drawing carried further than it is at present before we can
either attain to, or deserve, much improvement in the
different kinds of the woollen industry. Why, then, should
there have been such a reckless haste in the adoption of
the condenser? Such a course of headlong heedlessness
would be inexplicable and beyond comprehension were it
not that it has passed into a proverb that

> " Evil is wrought by want of thought
> As well as want of heart."

Surely nothing less than utter want of thought could
ever have led to such an extensive use of the condenser,
and to such an uprooting of the best interests of the
woollen manufacture. Not only has there been want of
thought in its adoption, but a further want of thought in
the extreme extent to which the condensing idea has been
carried. When the condenser first came into use the
rings on the doffer were not less than an inch in width ;
now they are under $\frac{3}{8}$ of an inch in many mills. If
the width of the ring was limited to 2 inches there
would be some chance of edging in a little drawing, but
the spinning-drawing never seems to come in for any consideration, or as being of any value at any turn, though it
is *the only thing that ought to be considered ;* everything else
is subservient, and only leads up to the spinning at this
stage of the manufacture. It is the only rightful possessor
of the ground and of the rights over which the manufacturing operations are passing at this point. When the
condenser comes to be clearly understood it will be found
that the machine is purely a usurper, that there is really

no place, no legal standing, in a manufacturing sense, for it in the woollen trade. It has done nothing but spread ruin in every possible direction by *cutting off the spinning-drawing*, and thereby ruining the manufacture. It has unfortunately stepped in and has "condensed" where "expansion" was most urgently needed. It has shortened the processes where it ought to have lengthened them, has done the very opposite of what ought to have been done, and at every step has spread blight and ruin wherever it has gone. It cuts off the drawing and leaves the thread only partially formed, deprives it of its *cover*, of its *strength*, of its *elasticity*, in fact, it deprives the woollen thread of *everything that makes it valuable* as a manufacturing tissue, and has done this by taking its life's blood—the drawing. And it is not only in woollen, but in every other thread. It is drawing that gives the manufacturing tissues their value, whether in point of elasticity, strength, suppleness, cover, or otherwise, as the case may be, woollen or worsted, as well as every other peculiarity or property that goes to make up the sum total of their value as textile tissues for manufacturing purposes. All their valuable properties are derived from drawing, and this the condenser destroys, as far as the woollen is concerned. In the instance cited, where noils have had to be resorted to in order to obtain cover, a better cover would have been obtained with much fewer noils, in the ordinary course of the structure of a true woollen thread, as well as every other property and peculiarity that makes the true woollen thread valuable.

Not long ago I saw a condenser at work turning off a twenty-four skeins sliver, intended to spin direct from the condenser to a little over fifty skeins. This is a very glaring instance of wantonly throwing away spinning-drawing when there is every opportunity of using it to much advantage ; the material could just as well have been condensed to ten skeins, then roved to twenty-four, and spun to the fifty odd skeins, and by this means the thread would have obtained five times as much drawing, and as a consequence immensely more cover, strength, elasticity, and manufacturing value.

The person that first devised such a procedure of condensing to twenty-four skeins and then spinning direct, at one operation of drawing, to upwards of fifty skeins could have no knowledge of spinning, or the value of spinning, or what effect spinning is intended to produce. If our forefathers of fifty years ago had had a material of this kind to make into fifty skeins yarn, they would have carded it to about 2½ skeins, slubbed it to twelve skeins, roved it to twenty-four skeins, and then spun it to the fifty odd skeins required. They would in this wise have put more than twenty times as much drawing into the thread than their overwise sons of the present day would put, and as a consequence they got proportionately more cover, strength, elasticity, and usefulness out of it. They were not so overwise as to think that they could get something out of the thread that they had never put in, but which ought to have been put in. If they wanted strength they knew that they could only obtain it by a certain arrangement of the fibres, side by side, and properly interlocked into each other's grip by a certain mode and amount of drawing. If they wanted cover on their thread they knew quite well that the material would need to be thoroughly well carded in the first place, and then that the cover could only be obtained by giving the amount of drawing required to eliminate and liberate each end of every fibre, and cause those ends of fibres to stand off from the core of the thread ; this never takes place in a condensed thread, only very partially, and never in any kind of partially-drawn thread, by whatsoever means produced, but only when the thread has been thoroughly drawn and every fibre arranged in its place, can the ends of the fibres rise to the surface in due course as the thread nears the completion of a thorough drawing. What amount of thread structure is it possible to get out of one unit of drawing obtained in spinning a twenty-four skeins sliver to about fifty skeins at one operation ? In a sliver so small as twenty-four skeins there is no body of fibre to lay hold of in order to begin to form it into a thread, and when the apology for the spinning operation has been finished the thread has nothing to stand by, only the name. Whether the thread is intended for the coarsest fancy woollens, any kind of middle-class cloth, or the finest west of England super, there is no woollen cloth of any kind where it is not a great advantage in every way to have a fully developed and complete thread at command for the purpose. Even if there has to be no milling employed in the manufacture of the fabric, and it is the very coarsest of fancy woollens, a good covered thread gives a genial handle to the cloth, and shows off the colours to the best advantage ; or should the yarn be made into a Bannockburn tweed, a fully-made woollen thread, in its very fullest cover, is of the first importance ; indeed, it is the primary condition of success, as it is a thread of this kind, very hard twined, that constitutes the foundation of the texture. If the tweed is to have a little milling, it will have to be taken out of the milling machine as soon as the end of the first stage is reached, while the thread is in its fullest swell ; for this also a fully-developed well-covered thread is of the first importance, as it gives a fulness to the cloth which nothing else can, and a mellowness not otherwise attainable. In any kind of middle-class cloth, cover of thread is a leading feature in getting a good handling fabric, and in bringing the colours out in their freshest and most advantageous aspect. A bare thread always gives a gray, washed-out, and worn appearance to any cloth.

And further, again, should the thread be intended for a plain super west of England cloth, nothing whatever can be done without a thread possessed of adequate cover, as nothing can supply its place or atone for its absence. Stripping the thread of its cover by adopting the condenser ruined the west of England manufacture. The cutting off of the greatest portion of the drawing by the adoption of the condenser was the cutting off of the only means of obtaining cover on the woollen thread. Without the labour of woollen-drawing no proper cover will arise on the woollen thread, and this only on the condition that the wool has previously been well and thoroughly carded; and without the cover obtained by drawing the production of a super west cloth is impossible. Fifty years ago, when so many of the productions were plain-faced goods, to such a degree were the manufacturers alive to having a full cover on the thread or yarn that when the thread had a glossy appearance, slightly resembling the gloss upon the present condenser thread, the carding engineer was told that he must card the wool a little farther, that he was leaving it short of work, and that he must work off the glaze, or work off the gloss, or the cloth would be what was then called "broken bottomed," meaning that the cloth would not have a full and even balk upon it, and consequently that the finisher would not be able to get a proper face upon the cloth in the finishing, nor be able to bring out to full effect the other properties of handle, mellowness, suppleness, etc., so characteristic of a true woollen manufacture. It was quite well understood in those days what would be the effect of using a bare, glossy, half-wrought, uncovered yarn. Without a well-carded and well-covered yarn the manufacturer knew that it was vain to expect that a well-covering and even-bodied balk could be got upon the cloth, knowing full well that unless a full even balk appeared upon the cloth he need not expect to attain that handle, suppleness, mellowness, and lustre of face which would be otherwise possible, as they all spring from the same foundation—a full-drawn thread. Whether our object is to produce a full-faced cloth or to leave the fabric in its balk state in respect to finish, still, in whatever state we decide to leave our fabric and consider it "finished," a full and thoroughly drawn thread gives us such a texture that we are left at full liberty to indulge our taste and fancy to the fullest extent, as we shall be able to get such a close and even-bodied balk upon it that we can leave it at almost any stage of the finishing process, and yet have a presentable and wearable fabric, suitable either to the roving and Bohemian tastes of the present day, or we can bestow a full-face finish upon it fit for the dress circle. The foundation of all true textile manufacturing is to have all the earlier operations so conducted that each shall lead up to and contribute towards the production of a fully developed thread, and upon this hinges all the issues as to whether the outcome of the manufacture is to be a success or a failure, whether we are to have a true cloth possessing all the desirable characteristic properties of strength, suppleness, handle, balk, and face finish, if we wish them. Quality and colour apart, all the other different characteristic properties of a truly made cloth arise from one and the same foundation—a fully drawn and fully covered thread. In a fully drawn thread the cover rises little by little as the drawing proceeds, and only as the drawing nears its completion does the cover near its completion ; both arrive at their full development together, at one and the same time, and by one and the same means. The reversal of the twine at each operation of drawing aids and helps very considerably in enabling the cover to rise, as a portion of the ends of the fibres get liberated at every reversal of the twine.

Woollen spinning has hitherto, at least during recent times, received very little attention, seeming to be unworthy of notice, and has been very little studied ; and the principle of its thread structure has received nothing like the amount of attention that has been given to the weaving ; but we have reached a point in the woollen manufacture when attention will have to be given to the material of which the pattern is formed, as beauty of form is only one thing, or one source of perfection in a structure. There must be something besides form, or the structure will have little more than the durability of a temporary shadow. There must henceforth be equal study and attention bestowed upon that which forms the pattern, the thread, as well as the pattern itself, as the perfection of the whole is composed of the perfection of the individual parts, and if this is not more closely attended to than it has been of late years we shall find that the manufacture has no more durability in it than a castle of cards. The want of attention to the thread structure, and all the earlier processes that lead up to it, coupled with the fearful falling back that has occurred during the last fifty years, or since the introduction of the condenser, has reduced the woollen industry to a most pitiable condition, not only in respect to the production of *plain goods* that are faced, but the deterioration has affected every class of goods, *plain or fancy*, to which a woollen thread can be applied. Any partially formed thing can only give off partial results, and the woollen industry has not only failed to improve its thread structure during the last fifty years to the extent that the worsted and cotton, as well as other industries, have improved theirs, but it has failed to stand still and maintain its ground; it has actually *fallen back* to an alarming extent, and the woollen thread of the present day has only about one-twentieth part of the build or structure in it that the woollen thread of fifty years ago possessed, without mentioning the fifty years' improvements that it ought to have acquired during the period to place it abreast of its twin industry, the worsted.

Looking these facts in the face, and taking them in all their bearings and consequences, can we be surprised that the woollen manufacture is not in that flourishing condition that it used to be in formerly? Is it not time that thread structure, as well as pattern and cloth structure, should be taught in our technical schools and colleges? In manufacturing there is something to be taken into account as well as the design—there is the proper structure and beauty of the material of which the pattern is to be composed, upon which will greatly depend its beauty and durability. A partially formed thread not only fails to be durable, but it fails to give full effect to the colour and beauty of the material, and in every way fails to bring out the full effect of which the material is capable, not only in its general effect on the cloth, but in the general effect upon any pattern or design to which it may be applied.

Taking into account the want of attention to thread

structure, and the loss of structure that has occurred through the introduction of the condenser during the last half-century, the marvel is not that the manufacture is down at a very low ebb, but that it has any foothold left, after labouring under such adverse influences for so lengthened a period. Surely the time is rapidly drawing near when the principles of thread structure in woollen spinning will get corresponding attention to that which is bestowed upon worsted and cotton. We will not undertake to say with what degree of consistency cotton spinning is taught in the technical school of a woollen district, while the spinning and thread structure of woollen is not taught, and while the district is mainly upheld by its woollen industry, and furthermore, where there is a lamentable want of all technical knowledge of the science and principles of the structure of the woollen thread and its spinning. The staple industry of any district ought to hold the first place in its technical school, whatever that industry may be ; and where woollen is the staple trade, woollen ought to have the first place, whether as to designing or any other department. Let each district do its utmost to place every department of its staple industry on the best scientific basis by teaching the principles involved in the various departments of its manufacture in the classes of its technical schools, and then technical schools will commend themselves to popular notice as being something better than ornaments, by becoming of practical use.

A further point to be noted is the reversal of the twist at each operation of drawing. Take the first operation of spinning, that of reducing the thick sliver, and what was formerly called *slubbing*. By the time that the stretch of drawing gets completed the fibres get locked fast in each other's embrace or grip, and before another stage or process of the spinning-drawing can be undertaken the twist must be reversed by setting the spindles to revolve in the opposite direction, so as to unlock or release the fibres from each other's grip, and set them at liberty to take part in a further operation of drawing called *roving;* and in this operation, as in the previous one, by the time that the mule has reached the end of the stretch or draw, the fibres have again become locked fast, as before, in each other's grip, and the drawing could not with advantage be continued longer, as the thread would not draw steadily and uniformly, but would be subject to sudden slips, and become very uneven and irregular in different parts of the stretch; so that the reversing of the twist or twine has to be again resorted to by setting the spindles to revolve in the contrary direction to that in which they revolved in roving. The twist having been again reversed, the third operation of the spinning-drawing is commenced, which is this time called the finished spinning.

In each one of these operations of spinning-drawing the arrangement of the fibres is advanced a stage towards that complete form or position in which they are required to lie in order to constitute a woollen thread; each operation of drawing with reversed twist facilitates and helps forward towards the completion of such arrangement; indeed, this reversion of the twist at each operation of drawing is an absolute necessity in the case of the woollen thread, and admits of no exception. Even in roller-drawn threads such as worsted and cotton it is found to be an improvement to reverse the twist at each operation of drawing. The twine going into the thread at one and the same time as that in which the draft is being effected, and forming part of the woollen process or method of drawing, there is a strong tendency in the fibres to get locked fast, and for the drawing to cease at certain portions of the thread after the drawing has been in operation for the space of about one-half of the distance of the length of the stretch of the mule; this is especially the case in finished spinning. In roving the drawing may be extended or prolonged to a greater distance before this locking becomes general all over the thread, and in the slubbing, or first operation of woollen spinning, the drawing may be extended to four-fifths of the whole stretch of the mule. There is not the same tendency to lock in roller-drawing, as the twine does not go into the thread during the drawing, but afterwards, whereas in woollen spinning it is the twine going into the thread at the same time that it is being drawn that arranges the fibres in that peculiar manner that constitutes the woollen as distinct from the worsted or roller-drawn thread, both being of identically the same wool.

The wool industry, in its two divisions of woollen and worsted, embodies both principles of spinning—the old and the new. The woollen embodies the old principle of spinning, and the worsted the new, or the principle of the roller-drawing of the Arkwright frame, or what was first put into practical working shape by Arkwright, so that the two textile threads composed of wool cover the whole ground of spinning, ancient and modern, and the student can sweep the whole range of spinning in the different applications of one and the same material—wool—to the textile manufactures. In this respect the student of the wool material as an article of manufacture is more advantageously situated for getting a thoroughly grounded knowledge of the different modes or principles of spinning than the student of cotton, flax, spun silk, etc., as he has the opportunity of clearly ascertaining what is the full effect of the different modes of treating or spinning the same material—wool, which knowledge he cannot gain as thoroughly and ground himself as completely in elementary knowledge of spinning in all its aspects and bearings from any other material of textile manufacture. Seeing the great difference that there is between the spun threads of worsted and woollen, the student will naturally feel more curious and interested to understand clearly what it is that constitutes the difference; the fact that both threads are frequently produced from identically the same wool, gives increasing zest and interest to the study of finding out wherein the difference lies, and leads to a thorough understanding of what it is that constitutes the difference between the two threads, showing that there is nothing on which to base the difference but the two opposite modes of converting wool into two distinctly different textile threads, the two quite opposite results resting solely upon the two distinctly opposite modes of procedure.

The student will be well repaid for a little extra labour and study at this point, as the whole of the general principles of spinning, both old and new, are illustrated in one and the same material—wool, which do not occur in any other material of textile manufacture.

This being the case, there is no evading the conclusion that the difference between the woollen and worsted threads arises entirely from the two different modes of preparing and spinning wool. With this conclusion settled and fixed firmly in the mind, the student is not so likely to be turned aside from making a thorough and detailed inquiry into everything that goes to make up the whole of each of the two principles of spinning that are involved in the two wool industries of woollen and worsted.

It is best to study the two methods of working or spinning wool in connection, as the one illustrates the other. In our study we no sooner find out a principle that is applicable and workable in one industry than we immediately find out that it will not work at all in the other. The modes of procedure followed out in producing woollen and worsted are as opposite as the poles, and cannot be made to approach or mingle with each other without doing violence to both. The two methods of working wool into woollen and worsted are utterly and irreconcilably antagonistic to each other, and each of the two textile threads of worsted and woollen is produced by two elaborate and opposite systems of drawing, which are brought out into greater prominence in the wool industries than in that of any other raw material of textile manufacture. In the two wool industries the student finds spinning stripped of all extraneous matter and narrowed down to a distinct and absolute principle, and if he fails to arrive at a satisfactory conclusion on the matter and thoroughly understand the whole subject, it can arise only from want of patient, searching investigation. No other material presents such an instance in illustration of the two principles of spinning as wool. We have seen that in spinning a good quality of 48s. worsted no less than *ten thousand millions* of crossings and re-drawings are involved in the operation, whereas in woollen there are no crossings of the sliver after it has left the carding machine. After the drawings commence in woollen there is no such thing done as putting two or more slivers together and drawing them into one. The woollen is essentially a one process manufacture, and if a fault is made in the carding or spinning processes, it remains a fault in all its nakedness to the end of the manufacture, while in worsted, if anything occurs to make a faulty or imperfect sliver, that imperfection can be toned down or distributed amongst a great number of other slivers, through the immense number of crossings that take place in the worsted drawings, till the fault becomes imperceptible; while, on the other hand, there are no crossings or putting of two or more slivers together in the woollen. Whatever the worsted *is* the woollen *is not;* the drawing in the one is the very antithesis of the other; one is full of crossings running up to thousands of millions, while the other has not a single crossing in its drawing, but is distinctly a single operation manufacture, as far as the spinning is concerned. In woollen spinning each operation has to stand on its own merits, and can never be crossed with another; therefore a greater amount of care and watchfulness is absolutely necessary in conducting each individual operation, as there are no points where a fault can be wiped out, as in worsted, cotton, flax, and other roller-drawn textile threads. This being the case, it is very necessary that the carding operation should be managed with the greatest care and skill, so that the cardings or thick slivers, from which the first thick spinning or slubbing is formed, shall be as uniform and perfect as is practicably possible, as any irregularity cannot be rectified in the succeeding operations. The woollen is distinctly and essentially a manufacture that hangs upon the perfection, so to speak, of individual processes; and where everything depends on the correctness of one operation, if that operation is imperfectly performed, the work is permanently injured and spoiled. Too much stress cannot be laid upon good and uniform carding in the woollen manufacture, so that the drawing may be commenced from as perfect cardings as it is possible to produce, which is an immense advantage at the outset of the drawing operation. There is not that absolute need of the same high degree of care in carding for worsted in respect to uniformity as for woollen, as the opportunity of crossing the slivers will hide faults that would be great drawbacks in woollen. Hence arises the necessity in the woollen for setting the cards in such a manner that one uniform sharpness of point shall be kept as far as possible week in and week out, and that the speeds shall be adjusted to the work in hand with such nicety that the material shall pass through the carding machine with the greatest possible order and regularity, fancies not running at too high a speed, so as to throw the material out of the machine and about the room and cause unevenness in the work, nor the middle and last doffers too slow, so as to cause the machine to work too thick in wool, and thereby turn off rough and irregular work. No other industry has to depend so much upon good and uniform carding as the woollen.

"Woollen yarn," says a recent writer, "has, until lately, been spun only on the mule since that machine was invented ; but a spinning frame, upon the throstle principle of continuous drafting, has lately been made, which is suitable for a sort of woollen yarn." This is a radical mistake, and betrays a confusion of ideas as to what is really worsted and what woollen, and what machines or action of the machines produce worsted and what woollen. The woollen thread does not admit a particle of spinning framework or continuous drafting in its structure; let this be distinctly and clearly understood at the outset of all study of woollen spinning. Instead of a thread drafted on a frame on the continuous principle being "*a sort of woollen yarn,*" it is just the opposite — it is a sort of worsted yarn. Things must incline in their very nature towards that to which they are related, and if the yarn in question is drawn on a frame, and on the continuous principle, it must of necessity be a sort of worsted. The continuous principle of spinning by roller-drawing belongs to the worsted and all the non-covering yarns, while the intermittent spindle-drawing principle of spinning belongs to the woollen and its allies. The two different principles of drawing constitute the dividing line between the two wool threads, worsted and woollen. Let us have a clearly defined scientific principle of production, and then we shall be able to classify the things produced. Continuous roller-drawing produces worsted, and intermittent spindle-drawing produces woollen.

The same writer's reference to the mule rather muddles than clears up the distinction between worsted and woollen. The mule is a machine of such immense range that it sweeps the whole horizon of spinning, both on the old principle and the new; it can spin either of the two wool threads or yarns, whether worsted or woollen; but in order to do this it has to be fitted up as a roller-drawing machine to spin the worsted, and as a spindle-drawing machine to spin the woollen. If behind its front rollers there are placed the series of drawing rollers of the water frame of Arkwright, then it is fitted up for spinning worsted, and, with modifications, cotton and the other roller-drawn yarns. But if, on the other hand, the mule is intended for woollen spinning, the series of drawing rollers behind the front row are dispensed with, as there is no roller-drawing in woollen. The mule can spin any kind of yarn, either on the new principle of roller-drawing, or on the old principle of spindle-drawing, roller-drawn yarns or the spindle-drawn yarns, according as it is fitted up to do the one kind or the other, or it can combine the two principles of spinning in almost any degree, as is frequently done more or less in the cotton manufacture, and as is sometimes done both on the Continent and in England in the finishing off of the French worsteds. The mule is what its name implies, a machine compounded of two natures, and can face in turn, so to speak, either to the north or to the south of the spinning world, so that when a mule is spoken of we require to be informed whether it is a worsted mule or a woollen mule, as their modes of action, or the principles on which they work, are the opposites of each other. The mule is, in fact, a universal spinning machine, and can take in either the old principle of spinning or the new, or it can combine the two principles; but there is such a strongly marked individuality about the woollen manufacture that it cannot admit of any mode of procedure other than its own into its processes without serious injury to itself. The woollen thread is a purely spindle-drawn tissue, and does not allow of one particle of roller-drawing of any kind in connection with its spinning, as it cannot be spun on any other than the intermittent principle of spindle-drawing.

The mule, when fitted up for cotton spinning, works intermittently, yet it is at the same time spinning its cotton on the continuous principle of the Arkwright roller-drawing frame, and the rollers are running during the whole or nearly the whole of the stretch or draw of the mule, and its action is only intermittent because it is a mule, and not a continuous acting throstle frame; the principle of its action, when at work on cotton, worsted, or any of the roller-drawn yarns, is that of the throstle frame or continuous spinning. While the mule is spinning woollen its action is purely that of the Hargreaves jenny, viz. that of a purely spindle-drawing machine. When the mule is used for the finishing part of worsted spinning it is fitted with the drawing rollers of the Arkwright frame, and works on the continuous drawing principle, though it is a mule, and the outward form of its action is intermittent; the real principle of its action is that of the continuous drafting spinning frame, with such modifications and improvements as the mule is capable of effecting. We require to be informed from which side the mule is working, whether from the roller-drawing side or from the spindle-drawing side, whether it is doing the work of a roller-drawing machine or the work of a spindle-drawing machine, before we can determine the character of the thread that it is producing; we require to know which side of the mule is taking the lead, the roller part or the spindle part, for each of the parts, as the one part or the other may be taking the lead, produces opposite results. As we have already said, when the mule is engaged in spinning woollen it does not use the drawing rollers, as woollen is drawn by the spindles and not by rollers. If mules are intended for cotton spinning, or for the finishing part of worsted spinning, they are fitted up with drawing rollers by the machine maker; but when the mules are intended for woollen spinning, the drawing rollers are left out and only the front row of rollers left in, as only one row of rollers is needed in a woollen spinning mule; the one set of rollers left in is not in any way or manner used for drawing, but only to wind in the sliver previous to the commencement of the drawing. We have said that the mule works intermittently, even when spinning cotton and worsted—when it is doing the work of the continuous spinning frame—but it only works intermittently because it is a mule and can work in no other manner, not that the work upon which it is engaged requires it; cotton and worsted admit of continuous drafting; indeed, continuous drafting is the constitutional principle of their respective threads. But when we turn to woollen spinning we are confronted with quite a different state of things; not only does the mule work intermittently, but the work of woollen spinning admits of no other than the intermittent mode of action; the nature of the work binds the operation down to this mode of action alone. In spinning woollen we wind in a length of sliver, say a yard in length, then stop the rollers and commence drawing by the spindles, and we must finish the drawing of this yard of sliver into a thread, and wind it upon the spindles and thoroughly dispose of it before any more sliver can be admitted, as by no other mode of action can the fibres of wool be arranged in the form required to constitute a woollen thread. The mule has not only formally to work intermittently, but it has actually and absolutely to do so in order to conform to the only mode known whereby wool fibres can be arranged into the form that constitutes a woollen thread. Were any sliver to be admitted by the rollers during the process of drawing by the spindles, the stretch of forming thread would be utterly spoiled, as it would stop the formative process that was going on amongst the fibres composing the draft, and the resulting tissue would be neither worsted nor woollen, but a mongrel cross between the two threads. The true woollen thread can only be drawn intermittently in stretches, one stretch at a time, and that must be taken in hand, completed, and disposed of before another length of sliver can be admitted.

The following extracts are from a series of articles which I contributed to the *Textile Manufacturer* in 1876 on "Textile Threads":—

"Modern spinning, on a large scale, may be said to

date back to the period of the invention of the mule by Crompton, or, at least, that particular phase or development of it which gave rise to our factory system. Prior to this period, for thousands of years spinning was a manual and domestic occupation or handicraft, practised with distaff and spindle, or with the one-thread spinning wheel. Shortly, however, before the period to which we allude (1779), Hargreaves, the pioneer of machine spinning, invented the 'jenny,' and Arkwright matured, not invented, the spinning frame. Hargreaves represented the one great system of intermittent spinning by spindle-drawing, and Arkwright represented the great system of continuous spinning by roller-drawing; the combination of the vital parts of the jenny spinning of Hargreaves and the roller spinning of Arkwright resulted in the production, by Crompton, of the 'mule,' a machine so unique and of such singular adaptability as the world had not hitherto seen, and such as it may not soon see surpassed. Such a leap forward did the new machine take in the production of fine yarns in the cotton that it was at first designated the 'muslin wheel.' But neither the most lissom fingers of the most subtle Hindoo, nor yet those famed and gifted damsels of the famed Asiatic city of Mosul, from which muslin is said to derive its name, with all their quickness and delicacy of finger-feeling could rival the productions of the 'new wheel.' The most fabulous prices were paid for the new muslin yarns, as much as 42s. per lb. being offered for No. 80s, and the new wheel soon scored the unprecedented number of 350 hanks per lb. amongst its achievements. The mule embodies in its structure all that we know of the principles of spinning. It is capable of being worked on the strictly woollen principle of the jenny of Hargreaves, or on the strictly roller principle of the water frame of Arkwright, or in any intermediate way between the two.   It can be applied to the spinning of all fibrous substances, to any kind of wool, vegetable or animal.

"Though in the cotton trade (for which it was first invented) its capabilities have been freely used, in other industries its capabilities have not been so freely used, nor so fully appreciated.   No sooner had the mule got to work than it was found to possess all the advantages of the water twist frame, in point of drawing power, without its disadvantages.   When the mule in the spinning of cotton had reached the point at which the frame was obliged to stop, it was found that the material could be carried forward in the spinning to a much finer count, as the radical and inherent defect of the frame was not found in the mule. In all frame spinning, no matter what the industry is, recourse is obliged to be had to the fly, or something equivalent to a fly; the continuous action of the frame does not admit of the fly being evaded.   The fly is an inherent defect in all frame spinning when applied to fine counts, and cannot be got rid of; either it, or something equivalent to it, must be used, in order to wind the finished thread upon the bobbin or spindle.   In the structure of the mule in front of the rollers, where the fly is used in the frame, the machine is composed of the best part of Hargreaves's jenny, and so avoids the necessity of the fly, and, instead of the sliver as it emerges from the front roller being burdened with the strain and pressure of the fly, it is left perfectly free from all strain or external pressure, having no strain or pressure to bear, but only has to sustain its own weight.   This freeing of the thread from all weight or strain during the process of spinning is one of the great achievements of the mule, seeing that it is attained without losing the advantages of roller-drawing. At this point, if we will only look thoughtfully into the matter, we shall see good and solid reasons why the mule should, once and for ever, leave the frame behind in all contests of fine spinning. For, whatever may be the highest achievements of the frame, the mule will always be able to write 'Excelsior' above it. The burden, strain, and external pressure of the fly in . all frame spinning must ever tend to break down the thread long before it has made any approach to the extreme tenuity attainable in the mule." The cotton and woollen industries have shown the greatest appreciation of the mule, and have been most disposed to allot to it its true place in the spinning of fibrous substances. In all industries where fineness of counts, finish, and extreme delicacy of manipulation are requisite, the mule is our most advanced machine.

A great number of frames have been brought out for the spinning of woollen yarn during the last forty years, but there is no framework in a true woollen thread. There are such machines as the " Vimont Spinner." " The continuous system of spinning" (says M. Charles Leroux) " is one of the great applications of mechanics to the treatment of carded wool (woollen yarns). For many years this problem seemed very difficult to solve ; the difficulties were so numerous that the constructors who tried to build on this principle were not rewarded with success."

" It was reserved for Mr. Augustus Vimont of Vire to transform the intermittent (mule) into the continuous frame (throstle). The industry of carded wool was the only one without its throstle frame ; we owe, therefore, this great progress to this skilful constructor." Frames for the spinning of woollen yarns have appeared in rapid succession during the last forty years. Like bubbles on

the stream they have appeared, things of a moment; then
they die, and yet the mania still possesses our mechanics
that the thing is to be done, ought to be done, *that it is
possible* to be done, whereas *a clearer and deeper* insight
into the nature of the thread to be operated upon would
convince them that success could not reasonably be expected in so barren a region. If the industry of carded
wool (woollen) " was the only one without its throstle
frame " it might have struck this French writer as being
somewhat singular, and that possibly there might be some
good and valid reason beneath the surface for this state
of things, and that there was no place for the frame
in the carded wool industry. Anyway, such is the
fact; there is no legitimate place for the frame in
the spinning department of the woollen industry. The
question of the application of the frame to the woollen
industry lies quite outside the domain of mechanics. The
Vimont frame exhibited in Paris in 1867 may be taken
as fairly representative of its class, but even the very
clever arrangement of its porcupines to aid the drawing
does not overcome the difficulty. If the matter at issue
was simply one of the few impossibilities that our wonderfully advanced mechanical age could possibly surmount,
there might be some hope, but the whole gist of the matter
lies outside the range of mechanical possibilities. It is
neither possible nor desirable. There is not a particle of
framework in the production of a pure woollen thread,
and, therefore, Mr. Vimont, in transforming the mule
into a throstle frame for the benefit of the carded wool
industry, has done nothing great, nothing more note-
worthy than the perpetration of a piece of mechanical
mischief. Ingenuity is worse than wasted in all such
feats. There are times in the history of all industries

when they diverge into wrong channels, and ramble off into by-paths, when it is necessary to hark back to the consideration of first principles, and the present we take to be one of those periods. In the carded wool (or woollen) industry, conducted on its own principles, there is no place for the throstle frame (or any other kind of frame) in its spinning department. "The carded wool industry," says this writer, "is the only one without its throstle frame." And very properly so, too; it is the only textile industry that has no place for the throstle frame in its spinning department, and ought it to be taken to task as having committed an unpardonable sin in doing without? Surely any industry has the right to discard or refuse that for which it has no legitimate use. Beginning at one end of the great series of textile threads, the spinning is all framework, and when the mule is working or employed at this end of the series it is working as an improved Arkwright throstle frame. Presently we advance farther into the series; we cross the borderland of pure frame spinning and enter upon yarns of that mixed class the production of which can only be accomplished on the mule.

The complex nature of the mule enables it to grapple with the production of a large class of yarns that are neither pure roller-drawn nor pure spindle-drawn, but which partake of a subtle combination of both systems of spinning; the wide sweep of capacity in the mule enables the cotton spinner of fine counts to use a combination of both systems of spinning in his yarns, and this remark applies to an indefinite extent to other industries, much more so than has yet been practically made use of. Any conceivable proportion of either one system of drawing or the other can be infused at pleasure into the operation as the process is carried on. The peculiarity of compound drawing is now being more generally understood and applied in the production of Angola and merino yarns.

Thus we may advance step by step through the entire series of textile threads, noticing as we pass what proportion of this yarn is the product of the Hargreaves side of the mule, and what proportion the work of the Arkwright side of the mule; in the other, finding at every step less and less of the Arkwright and more and more of the Hargreaves side, until we emerge at the opposite end of the series, where we encounter the carded wool, or woollen thread, which is entirely the product of the Hargreaves side of the mule, and which has not a particle of framework in its structure and nature. The frame cannot produce a real woollen thread; the Hargreaves jenny can; the mule can do the work of both, and people who are seeking to produce the woollen thread in the frame are but following a mechanical *ignis fatuus* that will lure them to their own destruction. We have described the structure of the woollen thread; we have not been airing opinions, but giving reasons and stating facts, and we have shown that the frame cannot produce a true woollen thread. It is not whether the French frame by Vimont is better or worse than some one of the English frames, but that spinning by the frame is bad in principle and false in practice when applied to the carded wool or woollen industry in particular.

From the extracts quoted we perceive that in taking a survey of the entire series of our textile yarns we can begin at one end of the series and find the thread structure composed entirely of roller-drawing, and then at the next step we find the thread to be less purely a roller-drawn one; and as we advance step by step we find the

threads become less and less purely roller-drawn, until
we find that spindle-drawing begins to predominate, and
that it predominates more and more as we advance further,
until we finally emerge at the other end of the series, where
roller-drawing disappears altogether, and it is to this last
link of the series that the woollen thread belongs, being a
purely spindle-drawn tissue, and as such it has to be produced by the Hargreaves jenny portion of the mule.

Spinning attained to its highest degree of perfection on the hand mule, as the draft was there capable of
the finest adjustment, and, as a consequence, the spinning
could be carried nearest perfection. It must be here understood that all the mechanical parts or movements involved
in what is, strictly speaking, spinning were complete in the
hand mule. The exact winding and measuring in of the
slubbing or sliver, and the exact mechanical and uniform
draft, along with the exact amount of twine within a given
space, all these essential points were complete, and were
brought out with the utmost mechanical exactitude in the
hand mule, and were capable of the greatest nicety of
adjustment. All the points of mechanical spinning were
complete in the hand mule, and the self-actor has done
nothing towards making spinning, as spinning, more perfect,
but rather the reverse, as the adjustment of the draft is
neither so easy, nor can it be so completely carried out as
in the hand mule. All that the self-actor has done, or all
that it can do, is to displace manual labour; it cannot
advance spinning any nearer perfection, as every movement
required to produce the thread was mechanical in the hand
mule, and it is only called the hand mule to distinguish it
from the self-actor, because at the production of every
stretch or draw of yarn manual labour was required to
wind up the yarn into the form of cops upon the spindle,

and to set the machine to work again to produce another stretch of yarn. Every movement required in the production of the thread of yarn was purely mechanical, and under much more complete control than they have hitherto been in the self-actor. Many people labour under the impression that in the hand mule the spinning somehow or other was partially done by hand, whereas the spinning or thread production was purely mechanical in every movement, manual labour only coming in to wind up the yarn and start the machine afresh at the close of the production of every stretch of yarn. We have already said, some pages back, that in the drawing of the woollen thread the twine was a fixed quantity, and that it was the speed of the draft that had to be varied so as to keep the twine and the draft on a balance while the thread was in the course of formation. The scroll of the hand mule could be much more readily adjusted to the requirements of the work in hand than the scroll of the self-actor. The bulk of the hand-mule scroll was composed of wood, the nose part only being of iron, and that only of very light construction compared with that of the self-actor, and any irregularity in the iron nose could be put right, or any desired alteration of form could soon be effected by a skilful spinner with the aid of a good file, but in respect to the ponderous scroll of the self-actor no such facility exists. There is too much sameness in the diameter of most of the self-actor scrolls in the nose portion. When we near the end of the nose portion of the self-actor scroll the diameter does not diminish with sufficient rapidity to meet the requirements of the spinner in very many instances. The skilful spinner and the skilful practical mechanic ought to lay their heads together, and set out an improved scroll for the self-actor mule. The improved scroll should be 16 to 18 inches

longer in the groove than the length of the stretch or draw
of the mule, so as to allow of the draft cord being withdrawn from the nose in certain cases, and to be let down
close to the extreme end of the nose in other cases, according to the exigencies of the draft in the varying materials
with which the spinner has to deal. If the self-actor
spinner had a scroll something after the kind here indicated,
with the loose thin circular segments previously mentioned at
his command, and which could be readily put into the bottom
of the grooves in the forepart of the scroll, he would be able
to adjust the draft much better than he can at present.
Some kinds of material that the spinner has to cope with
do not require the draft-cord to run down very far on the
nose of the scroll, while other kinds of material, especially
in the spinning of very fine counts, require the cord to be
run down to the extreme end of a rapidly diminishing
nose. These little alterations are very much needed to
enable the self-actor mule to come up to the standard
reached by the hand mule in the art and science of
mechanical spinning. Both the hand mule and the self-
actor mule are but tools in the hands of the spinner, and
the self-actor, for reasons already given, has not yet been
"improved" into as efficient a tool, even in the hands of
the most skilful workman, as the hand mule had become
in its later days. The ill-formed draft-scroll is a great
stumbling-block in the way of good spinning in the self-
actor. The extra strength that it has been deemed necessary to impart to every portion of the mule in making it
self-acting has made the draft-scroll into a very unwieldy,
ungainly thing, and in its transformation it has lost much
of its true form, through being left too much in the hands
of the mechanic, instead of being the joint product of the
skilful spinner and the clever mechanic.

The draft, and all that in any way pertains to it, is the most important section of woollen spinning, and everything that can be done ought to be done to make this thoroughly understood, to forward its more efficient accomplishment. It was one of the great achievements of the mule from the time that it was first invented that the draft became mechanical, and consequently more uniform in its character than was possible on the hand jenny, and that being once skilfully set, remained uniformly good to the end of the work in hand, unless through some carelessness in the carding department the work underwent some change, so as to cause the draft to have to be revised or readjusted to meet the change that had taken place through careless inattention, or through lack of adequate judgment in the setting of the carding machines in such a way that they should be able to turn off their work in one uniform manner from day to day. Seeing that the draft is purely mechanical, and so dependent on the skill of the spinner for correct setting, alike in the hand mule and in the self-actor, it is only reasonable that the draft should receive the greatest portion of the attention of the spinner, other things being of minor importance compared with this. There are many things about a self-actor mule of varying degrees of importance, but none of them of equal importance to the draft. Some spinners pride themselves immensely upon the form of the cop or bobbin that they can produce on the self-actor, and at the same time pay slovenly attention to the draft, forgetting that it is the draft alone that has the most abiding effect on the manufactured goods, as it has to do with the correct structure or otherwise of the threads composing the fabric, and the effect is abiding as long as the cloth continues to wear; whereas the form of the cop or bobbin is only a form, and

in this case is only of fleeting importance, and if not correct
in every particular, according to Euclid, it is of little
practical or enduring consequence, as it vanishes and is for
ever done with before the material gets put into the form
of cloth. The draft, therefore, ought to have the first and
the greatest portion of the attention, and ought to be most
carefully studied by all concerned in its most minute particulars. Nothing pertaining to the draft is unimportant
or unworthy of the closest attention. The spinner, in
particular, who is ambitious to become a good workman
should learn to discriminate between that which is of minor
and that which is of major importance in the practice of
his daily occupation, and to distribute his attention accordingly. He should not only give great attention to the
draft, but he should make it a long and very careful study.
Every change in the material, even the slightest, necessitates
some change in the draft. A younger wool, a little more
length in the staple, or greater pliability and softness, more
or less oil used upon it, different colour of dye, or difference
between spinning in the gray and in the dyed state, each
one of these particulars calls forth some change in the draft,
even providing that the quality in every instance is identically the same. These changes do not occur in any other
textile industry to anything like the same extent, but the
woollen-spinning draft is such a peculiar process, and is so
extremely sensitive, on account of its complex character,
that the slightest change in material affects it. We have
seen that the draft is not a simple process, but that it is a
compound process, arising from the twine pouring in and
attempting, as it were, to stop the draft by locking fast the
fibres in the grip of its gird, while the receding spindles are
attenuating the mass of fibres at the same time, and doing
the utmost that can be done to prevent the fibres from

getting locked fast by the twine. We thus see that the spindle is performing a double part of opposite tendencies, and it is this striving and contention between these two opposite forces of the recession of the spindles and the twine, each force striving to obtain the sole and absolute control, and to oust the other from the field, that the tediousness and the difficulty of the woollen draft arises. The woollen thread cannot be formed without the twine from the spindle, and it cannot be formed without the spindle receding from the winding-in rollers and stretching out and attenuating the fibres to a greater length; and then, again, the woollen thread cannot be formed without both these operations going on at one and the same time, and in opposition to each other, the twine striving to lock fast the fibres, and the receding spindle striving to prevent the fibres from getting locked fast by pulling them out into a greater length of tissue, and thus preventing the twine from having as great an influence upon the fibres as it would have if they remained the same bulk in diameter; so as the twine approaches the point, and is just about to accomplish its object in locking the fibres fast in its grasp, just as quickly the receding spindle pulls the fibres partly loose again, and only allows them to become parti- ally fastened in the grasp of the twine. Still the twine follows up with relentless persistency its attempts to lock the fibres fast in its grasp, and still the receding spindle, with equal doggedness and persistency, resents the encroach- ments of the twine and stretches out the fibres and pulls them out of the grasp of the twine, both forces struggling for the mastery, but neither the one nor the other being able to get the mastery over its opponent, or in any way to arrogate to itself any more than an equal share of power. Thus we see that the woollen thread is formed in the raging

war between these two forces (the twine and the recession
of the spindle), the twine persistently attempting to close
the breach, and the recession of the spindle just as persistently keeping it open, and so by the careful watching and
the skilful management of the clever spinner in keeping
these opposite and contending forces equally at bay, in
respect to each other, and allowing neither the one nor the
other to prevail over its opponent, but by keeping their
contentions strictly and equitably balanced, the woollen
thread is produced; and though it is the oldest known
system of spinning, yet the manner in which it is carried
out mechanically in our modern mills at the present
time makes it a subject of never-failing interest, and
one well worthy of all the attention that can be bestowed
upon it, with a view to bringing it to mechanical perfection.
Indeed it calls forth the best skill of the spinner in
playing off these two opposite forces against each other,
and in effecting the continuous balance of the forces with
the greatest nicety and precision of mechanical adjustment.
Ever since the early years of the invention of the mule the
woollen draft has been purely mechanical, as the draft-
scroll came into being with the early mule, and all through
the years of the existence of the so-called "hand mule"
the woollen draft has been as purely mechanical as it is
possible for it to be, and has reached a higher standard of
perfection on the so-called hand mule than has yet been
reached on the self-actor, for reasons that we have just
given, and we must not expect that we shall be able to
get a more perfect draft from the latter than we have been
able to get from the former; both are alike purely mechanical
in the draft, and both are capable of the same mechanical
exactitude, an exactitude that could never be reached on
the Hargreaves hand jenny. A good woollen draft is a

most tedious and difficult thing fully to comprehend, on
account of the complexity of its nature, and it is an equally
difficult thing to accomplish thoroughly well.   The draft is
a continuous balancing of the effects of two opposite forces
that are continually struggling to get free from control and
to get off the balance that is imposed upon them in respect
to each other, and they would get off were they not followed,
inch by inch, throughout the course of the entire length of
the stretch or draw of the mule, and prevented by an inch
by inch adjustment of the balance, the draft frequently
commencing at a speed seven times as great as that at
which it finishes the stretch of yarn.   The speed at which
the spindle revolves is constant and uniform, but the speed
at which it recedes is ever varying, and diminishes in the
ratio of from 7 down to 1.   This necessitates the latter half
of the draft-scroll being somewhat in the form of a cone.

It is this complexity of the woollen draft, and the effect
that this complexity has upon the arrangement of the fibres,
the two forces in antagonism, and the delicate balancing of
each against the other, that causes the woollen spinning to
differ from that of most other textile industries in its
spinning department.   In almost every other large textile
industry the basis of its spinning is roller-drawing, with
only a very slight portion of spindle-drawing added, as in
cotton, in that part of the draft that is technically called
the "gain," which is simply that additional speed at which
the spindle carriage moves over and above the speed of the
front rollers ; and from this explanation it will be perceived
that "gain " is not pure spindle-drawing, as the rollers are
all the time delivering or letting in additional sliver while
the "gain " is going on, which is contrary to the principle
of pure woollen spinning.   The draft part of the self-actor
woollen mule ought therefore to have much closer attention

bestowed upon it, both on the part of the spinner and on the part of the machine maker, in affording much greater facilities for the more delicate adjustment and setting of the draft than we have hitherto had. A longer length of scroll groove, something like 18 inches longer than the draw length of the mule, a more rapid and uniform diminution of the diameter of the draft-scroll during the last foot in length or so of the groove as it approaches the extreme end of the nose, and some kind of half circle and quarter circle segments of different thickness that could readily be put in and fastened down into the bottom of the grooves, with set screws for the forepart of the scroll, are imperatively needed before we can reach with the self-actor the present known standard of good spinning. The most important point, and that to which we ought to give the most careful attention, is the draft; other points are of minor importance compared with this. It is the draft that decides the value of the thread from a manufacturing point of view, and if the draft is done in a slovenly manner, and carelessly dealt with, the manufacture to which the work in hand may be applied suffers proportionately. Other portions of the self-actor mule have received their due share of attention, such, for instance, as the cop shaping, but the form of the cop is only of minor importance, as we have already shown, compared with the drafting of the thread, as the form of the cop or bobbin vanishes as soon as the yarn is put into use in the next stage of the manufacturing process, but the effect of good drafting remains as long as a shred of the fabric manufactured from it endures. Not that we would by any means ignore or despise a well-shaped cop or bobbin, but its importance is of secondary consideration compared with a good draft and everything that pertains to its improvement; so much so ought this to be the case that the chief attention should be concentrated upon it. The effect of the one is abiding, while the effect of the other is only temporary. The value of each ought to be decided on its merits, and their relative importance ought to decide the relative share of attention that each is to receive. There is yet much to be done to bring up the facilities for adjusting the draft in the self-actor to the standard attainable in the old hand mule. The mechanical system of spinning began really with the invention of the Hargreaves jenny, as it was a method of spinning a number of threads at once, but in the Hargreaves jenny the draft was not mechanical, nor was the draft mechanical in the early years of the mule. The mule was even a manual machine, as we have previously said, up to about the year 1790, and it was about this time that James Hargreaves (not the inventor of the hand jenny) brought out the more perfect method of bringing out the spindle carriage by a parallel scroll for the first half of the draft, and a conical scroll attached to the end of the parallel one for the latter half of the draft, and here we have the germ of a perfect and purely mechanical draft. Previous to this period both the twine and the drawing out of the spindle carriage had been purely hand work, and had been subject to irregularities in the hands of the best workmen. Sometimes the twine would vary and be insufficient in quantity for the ordinary speed of the receding carriage, while at other times it would be in excess, and then again alternately one and then the other during one and the same stretch of yarn. On the other hand, when the twine was tolerably constant for a time the speed of the carriage was liable to the same irregularities in its movements as the twine. In the same stretch or draw of yarn the speed of the carriage would be in excess of the twine, and thereby let the thread loose and sometimes break down, while in other stretches the speed of the carriage would lag behind, and thereby let the thread get set too fast by the twine, while again at other times both incidents would occur during one and the same stretch or draw of yarn. The best spinners could not at times avoid these occurrences, in a greater or a lesser degree, either in the twine or in the speed of the carriage; it could not in the very nature of things be otherwise. It is quite true that some spinners attained to a high degree of proficiency in the regularity of their step in bringing out the carriage, but it sometimes happened that the spinner that was most constant in his step with the carriage was not always the most constant in turning in the wheel and producing the twine and *vice versâ*; so that any deviations from uniformity, whether in the step with the carriage or with the turning of the wheel, spoiled the perfection of the draft. It was not until the twine became constant by becoming mechanical, and the invention of the draft-scroll, that spinning reached the stage of mechanical exactness, or to a stage where a steady balance could be maintained between the speed of the carriage and the speed or quantity of the twine all through the stretch alike, from the stoppage of the rollers to the finish of the draw of the mule. But ever since steam and water-power have been applied to the working of the mule, twine has been a constant quantity, and could be relied upon; and ever since the invention of the draft-scroll there has been uniformity in the varying speed of the receding spindle carriage that could always be calculated and relied upon from stretch to stretch all the day through, which previously had been very uncertain, but by both movements of the draft being placed under mechanical control a degree of exactness and uniformity was reached that had before been unattainable;

and as the speed of the twine and the speed of the receding spindle carriage are the two vital elements of woollen spinning, we perceive that from this time spinning became a purely mechanical operation, and as such was capable of a high degree of exactness in the way of uniformity between one stretch and another after the draft had once been properly set. This being the case, we see that the mule has been a purely mechanical spinner during the last hundred years (or ever since the invention of the draft-scroll), and the application of mechanical movement to the rim-shaft of the mule, from which both the spindles and the carriage derive their action, and any variation in the prime mover in respect to speed, whether steam or water-power, does not disturb the relative speeds of the spindle and the carriage, as both are propelled by the rim-shaft, and whether the rim-shaft revolves a little quicker or a little slower than the normal speed, the relative speeds of the spindle and the carriage remain undisturbed.

Since the introduction of the self-actor we have got into the habit of calling the old mule the hand mule, but it is a complete misnomer to call it so, as the mule has been a mechanical spinner ever since mechanical power has been applied to the operating of the two elementary movements that are vital in the structure of the woollen thread draft, viz. the mechanical constancy of the twine and the mechanical constancy of variation in the speed of the receding carriage. These two are the vital elements involved in the structure of a true woollen thread, operating in conjunction and evenly balanced against each other, as we have already pointed out, and these have been mechanical during nearly the whole of the years that the mule has existed. The manual labour that was used in working the mule was used in such a way as never to touch the mechanical action that was vital to the true structure of the thread. These since the invention of the draft-scroll, we have seen, have been purely mechanical, and capable of the most exact uniformity and the greatest nicety of adjustment, and afforded the same facilities for attaining to as high a degree of perfection a century ago as now, since the vital parts of the drafts were as strictly mechanical then as now, and the manual labour employed did not in any way trench upon the mechanical accuracy of the draft, and we are strongly inclined to think that the woollen draft was much better understood and practised fifty years ago than it is now. There is such an amount of mechanism about the modern mule that the spinner or man in charge gets lost in the maze of it, and does not sufficiently discriminate between what is *important and vital* and that which is only subsidiary or of secondary importance. The manual labour was employed in winding up the spun thread into cops or bobbins, as the case might be, but the spinning of the thread itself was purely mechanical, and therefore capable of being produced with the greatest possible uniformity and all the niceties of adjustment that the present day can boast of. It is a great mistake to suppose that because the minor movements of the mule have become mechanical in the self-actor that the quality of spinning will, as a consequence, be capable of great improvement. Such is not the fact, as we have already seen that the vital part of spinning (the mechanical draft) has long been in existence in the old mule, and that the possibilities of the highest excellence have been open to the skill of the spinner from the first days of the mechanical draft-scroll and the uniform mechanical twine. The self-actor opens out no additional facilities for a higher class of spinning than the old mule afforded.

We must look to a higher technical knowledge, and a fuller appreciation of the facilities we already possess, and make a better use of them as stepping-stones towards higher attainments. How much higher we can push the adjustment of the draft, and how much more delicately we can balance the two forces that constitute the woollen draft, a more extended technical study, combined with a higher practical skill, alone can determine. We have mentioned three separate and distinct drawings as being necessary to constitute a fairly good woollen thread in a good class of material, but we only limit ourselves to the naming of three as a present practical stopping-place, and one that is as advanced as the woollen manufacturing community can at present be induced to give ear to. Many manufacturers will demur to as many as three drawings, even for high-class goods, especially those who are condenser blinded, and have very rarely seen more than one drawing used. First-class work ought never to be allowed, *on any account*, to pass through the mill with less than three drawings. Second-class work, or medium, might at present be allowed to pass with only two drawings. The sister-wool industry, as we have seen in the case cited in respect to 48s. worsted, uses fourteen separate and distinct drawings, and if the best class of woollen manufacturers could be induced to bestow more than three drawings upon the best class of the production of their first-class yarns, they would find their account in it, as the worsted people have done, as the more drawing that is bestowed upon any drawn thread used in any textile manufacture, the more perfect in its kind that thread becomes; and if the other wool industry finds it an advantage to bestow fourteen distinct drawings upon its thread, combined with crossings that run up to ten thousand millions, thereby

carrying its development to such a high degree of perfection, surely the day will yet come when technical knowledge of woollen spinning shall have become more widely
diffused, and the complexity of its draft shall have become
better understood and appreciated as producing a textile
thread the most valuable to man, and one that contributes
most to his health, convenience, and comfort. When this
day of advanced technical knowledge comes, then woollen
spinning will be seen to be of at least as great importance
as worsted spinning, and will receive its due share of
consideration, and will have conceded to it the necessary
amount of drawing without its being begrudged as so much
unnecessary and wasted labour, but every necessary attention, thought, care, and skill will be called into its service.
During recent years woollen spinning has received scant
attention, in fact, it has been most shamefully neglected.
But when the complex structure of the woollen thread is
better understood, the beauty, utility, and durability of the
woollen arrangement of the wool fibre will be much more
highly appreciated, as the woollen thread presents the
fibres of wool in the most serviceable form for the structure
of warm clothing fabrics, whether for light underwear or
for the heavier purposes of outside wear. The better the
woollen thread structure is understood the more extensive
will its use become, for this reason, that when it gets to be
thoroughly and technically understood in every minute
particular, it will be found that its structure is such that
it presents the wool fibre in such a form as that when
developed to its fullest extent, in its woollen form, as a
textile tissue or thread, it is capable of an almost endless
variety of uses, and the more fully it is developed the
wider becomes its range of use. The woollen thread is in
a very low state of development at the present day, but

when its structure and use are taken up as subjects of study for the classes in our technical schools and colleges, just as its structure and use become more technically and scientifically understood and more widely known, in the same proportion will new fields open out for its application and extended use; these, again, will further extend and beget further. As wider diffused knowledge of the uniqueness and worth of the wool fibre in this form of tissue gets more and more studied and extensively known, the more will it be appreciated, valued, and used. Wool is one of the most ancient of the clothing materials with which mankind have any acquaintance, and because it is old and its manufacture familiar some people think that they know all about it, and allow newer materials to absorb their attention. But they do not know all about wool, and they do not know all its capabilities taken in connection with the mechanical advancement of modern times. Wool, like other textile materials, can be wrought up into textile tissues to a much greater nicety and perfection with the present mechanical facilities than was the case in former times, and so far from its uses being exhausted and worn out, we are only, as it were, at the beginning of the uses of a true and fully-developed woollen thread as it will be produced under the advanced mechanical, technical, and scientific manipulation that the facilities of our modern mill affords when rightly understood. With the exception of the application of pattern to woollen cloths in connection with the fancy woollen industry, the woollen manufacture has stood still, *comparatively*, during the last half-century, but when the earlier stages of its manufacture obtain their proper place and consideration we may expect to see some move in the direction of the woollen manufacture beginning to assume its rank again amongst

the great industries of the empire, and see it rise again
to its wonted eminence. The structure of the woollen
thread is such that it requires a very high degree of
mechanical exactness and precision to carry its development to its highest degree of perfection. The spinning of
woollen is a very fitting and a very pressing subject
for the study of our young men who are intending to take
any active part in the manufacture to which it belongs.
The woollen thread structure (judging from the little
notice that it has received during recent years) is fast
tending towards becoming one of the lost arts, and all
sorts of wild theories and crude notions are afloat, not only
as to how it ought to be produced, but even as to what
really is its nature and structure. It is the object of this
text-book not only to induce the student to strive to lay
hold of the principle of the woollen thread structure, but
to study the most scientific and practical mode of producing it. It has been wittily but tauntingly said that
"science is a first-rate piece of furniture for a man's
upper chamber, if he has common sense on the ground
floor." We have no disposition to disturb common sense
in its possession of the ground floor, but every disposition
to strengthen, fortify, and uphold it, and persuade science
to descend from its lofty heights and upper chambers and
assist common sense and the highest practical skill in
improving the ground-floor occupations and industries of
daily life. When woollen spinning and woollen thread
structure again get their due share of attention, and
become better known and understood than they are at
the present day, there will be no such thing as woollen
spinning at one operation of drawing, or such as is now
known as "spinning direct from the condenser," except for
yarns that are the lowest of the low. A low medium class

of yarns will then get two operations of drawing and the better class of yarns will get three or more drawings of a superior kind to what is now common, as the high mechanical exactness that is attainable at the present day will be more and more called into the service of woollen spinning, and the woollen draft will become more perfect as it gets more studied and becomes subjected to higher and more exact mechanical treatment and superior practical skill, directed by the highest technical and scientific knowledge.

We have already said that the object of carding is to break up all the natural arrangement of the fibres preparatory to their being arranged artificially into textile tissues, and if our worsted friends, taking the wool from the machine in its chaotic state, require fourteen drawings, with their innumerable crossings, to compel the fibres to lie down flat and parallel in order to get the utmost number of fibre ends embedded in the body of the thread, can we in the woollen manufacture reasonably expect to do the reverse of the worsted by one miserable half unit of drawing in spinning direct after the condenser? To dream for an instant of being able to do such a thing is the essence of absurdity. The worsted people can only compel the fibres to lie down by a long and laborious series of drawings, and we can only by an opposite and laborious system of drawings compel the fibre ends to rise up and stand off from the main body or core of the thread. One operation of drawing will not compel a tithe of the number of fibre ends to rise; the twine requires to be reversed and the drawing repeated several times before the fibres can be arranged in proper woollen form, and before any considerable proportion of the fibre ends can come to the surface, and it is only by a laborious system of drawings that the woollen thread can be produced in perfection. If we will still persist in dispensing with the drawing to which our forefathers had reached, we must prepare ourselves to endure the consequences. If we take the carded wool in the chaotic state into which it has purposely been thrown by the carding machine, and refuse to put the requisite labour into it in order to produce a proper woollen thread, we must not expect to get the service or value of a proper woollen thread out of it. What costs nothing is worth just as much as it costs; labour must be put in before value can be got out. Without drawing we can have no woollen thread of any value.

Can any contrast be more startling than the present-day practical dealing with the two textile threads producible from wool? Every requisite means are taken, on the one hand, to compel the fibre to lie down parallel and form a clean bare thread, like drawn wire, while, on the other hand, we in the woollen have gradually dwindled down to one half unit of drawing in spinning direct from the condenser, and with unblushing cheek call such a tissue a woollen thread, when even our forefathers, in their dark days of no scientific knowledge and no technical schools, had groped their way up through the darkness by which they were surrounded to twenty times as much drawing in their woollen thread as we bestow upon ours in these days of boasted light and knowledge, and with all the facilities of our modern mills. They lived in dark days, with nothing but common sense to guide them; we live in days of light, with technical and scientific knowledge around us on every hand, and coming to our doors and into our houses by means of the press, the platform, reading-rooms, and free libraries; indeed, light and knowledge are now so common that they come to us in the very air we

breathe; yet what use have we, in the earlier stages of
the woollen industry, made of all the light and knowledge, technical and scientific, that surround us? Of what
use is technical and scientific knowledge if not pressed into
service to improve everyday life, to improve the very industries in which most of us are daily engaged, and prompt
us to push both the industries and ourselves onwards and
upwards? Viewed from this standpoint, what has all the
light of technical and scientific knowledge done to advance
the early stages of the woollen industry? We scour our
wools with fixed alkalies instead of with volatile ones as
formerly. We steep our wools in bulk, and even manufactured pieces in sulphuric acid, to "extract the burr and other
vegetable matter," instead of sorting out the most burry
portions and "extracting" them only (if we must "extract"
at all) and allowing the "burring machine" to do the rest.
The burring machine is one of the most useful machines ever
introduced into the woollen manufacture, but instead of
using it we fly to chemicals and "extract" the burr forsooth, and at the same time destroy the working and
wearing properties of the wool. We have reduced spinning
to a mere pretence and a sham by *condensing* the material
instead of drawing it, when it is by drawing only in the
spinning that we can produce a thread.

As yet we have made no use of technical and scientific
knowledge in the earlier stage of the woollen industry, or
there would not have been all this going backwards in the
scouring of our wools and in extracting. Technical and
scientific knowledge would have taught us that it was
the height of folly to do violence to material on which we
are engaged. Verily, "a little learning is a dangerous thing,"
and it was never more strikingly illustrated than in the
backward movements we have made in recent times in the

earlier operations of the woollen manufacture. Scientific knowledge can tell us that we can scour wool by using fixed alkalies, but it can tell us something more. Scientific knowledge can tell us that we can destroy the "burr," or any other vegetable matter that may be found in a fleece of wool, by saturating it with sulphuric acid and then subjecting it to great heat; but it can tell us something more. Scientific and technical knowledge can tell us that we can produce a weavable tissue by condensing the material down to a very fine sliver, instead of reducing the thick sliver down by spinning; but it can also tell us something, very much of something, more besides, and this additional knowledge ought to be given to students in the class-room. Shall we in the woollen industry be the last to awake and avail ourselves of the advantages, and see the necessities, of modern times? The worsted people have advanced their thread structure during the last half century to the highest degree of practical perfection; we in the woollen have gone backwards. Ought we to regard so listlessly the interests of such a great, important, and ancient industry — an industry that traverses the long ages of history? My experience in connection with woollen spinning dates sufficiently far back to enable me to say as a practical spinner that the woollen thread used to get twenty times as much drawing as it gets now under the reign of the condenser. If we will still thoughtlessly or ignorantly reduce the sliver, contrary to all common sense and all science, by the condenser, instead of by the true method of repeated spindle-drawings, we reduce spinning to a farce, and utterly defeat the production of a woollen thread, which either is, or ought to be, the object of woollen spinning. There is no other way, as we have repeatedly said, of producing a complete woollen thread than that of reducing the thick sliver by repeated spindle-drawings, as we have already shown, and have given the why and the wherefore for such an emphatic statement. *Condensing* where *expanding* was so urgently needed stands to neither common sense nor science, but is solely and certainly the direct road to ruin in any industry, walk, or occupation of life. We have so far lost ground by losing sight of all principle in regard to spinning, that we require to begin *de novo*, as the Latinists say, and teach the elements of spinning and its relation to thread structure, and the bearing of thread structure upon the success of our manufacture, as we are fast losing all knowledge of spinning, its purport, place, and power in the economy of the woollen manufacture. Such being the case, there is no help for us out of our present position but to teach the elements of the early processes of our manufacture in our technical schools and colleges, so that the rising generation, and those of older growth, may obtain some elementary knowledge of the principles that underlie the various processes of the manufacture. So far is the woollen manufacture from being played out, that it is only in its rudimentary state, as will be seen when its peculiar thread structure is studied and the uniqueness of that structure becomes thoroughly understood and systematically taught. Then it will be seen that the possibilities of the woollen manufacture in the future are great and many, and that with a fully constructed thread the makes of old cloths will be revived *and improved*, and new makes will spring up with the exigencies of the times, and a thousand applications will be found for the completely developed thread (which its present shadow cannot fill), according to its almost boundless capabilities of employment and use.

Our thread structure has deteriorated to such an extent

that we cannot make or produce even a fancy woollen that will bear any degree of finish. The most popular finish of the day is that which has the least finish in it; the cloth is taken in its balk state and passed through the cutting machine, very open set, just to take off the ends of the long straggling hairs, and then it is ready for the press, and very properly so too, for the thread of the fabric is so raw, bony, and bare, that if the balk is disturbed in the least by any finish, its bony nakedness and rawness become so apparent that they are absolutely repulsive both to the eye and to the touch. The little bit of balk that accumulates on the surface helps to soften the colours, and enables them to blend and to mellow each other; therefore just take off the long ends of the fibres and send the cloth to press—this is the strait we are reduced to. Whereas if the fancy cloth had been produced from a fully constructed thread, full of fringy covering fibre, the colours would have mingled hair for hair, and softened each other's effect and toned down all abruptness of cut off and rawness of effect by mingling and mellowing each other's tones, and the cloth would have borne raising and cutting to any degree in the finishing that taste or the pattern might require without the cloth becoming repulsive either to the eye or to the hand.

And in cloths where there is no fancy pattern a fully constructed thread produces an effect both to the eye and to the hand infinitely superior to one produced from our present thread, no matter whether the cloth has to bear any finish or has to be left entirely in the balk. The superiority of the appearance will be strikingly manifest compared with anything that can be produced from our present very imperfectly constructed yarn. Our present imperfectly constructed yarn gives a grayness to even the best of colour, whereas a fully constructed thread has such an immense number of fibres standing off from the core at almost right angles, that the *ends* and not the *sides* of the fibres are presented to the eye, and we see little or nothing of the grayness which the sides of the fibres reflect, and in addition to this a fully constructed thread throws up so much greater a number of fibres which are all colour-bearers, and presents so much more colour to the eye, and presents it in so much better form.

Then take another illustration from the old plain cloth of two generations ago, which was full milled, and the garments made from it had to be handed down from one generation to another to finish off in wear. To construct such a cloth as the old plain used to be, the yarn had to have twenty times as much spinning-drawing as our present condenser yarn, and the result was that there was twenty times as much cover on the thread, and twenty times as much opportunity of getting proper balk on the cloth for covering the make or thread texture, and for obtaining a proper surface on which to get up a full-faced or any other kind of finish. And it must be clearly understood that the balk on the old superfine plain cloth was not a felting that went completely through the cloth, but that it was a proper balk, and composed only of the felting of the fibres that were on the surface of the cloth, both back and face. The compact texture of the cloth prevented the balk penetrating through the fabric, and confined the effects of the felting or milling operation more especially to the surface. We obtained all the strength that the well-compacted texture gave us when made from the comparatively well-constructed yarn of the period, and we had the suppleness which such a yarn affords us by the balk being confined to the surface, and that durability of wear and capability of finish that such a balk arising from such a texture was enabled to give us. What these were are too well known to need repeating here. If we could get such balk, finish, suppleness, and durability of wear from the yarns of fifty years ago, what may we not hope to get from a yarn that has 50 to 100 per cent more added to the spinning-drawing of the old yarn? The capabilities of such a yarn, from a manufacturing point of view, are past estimating, whether applied to goods of the highest finish or to goods with no finish at all upon them; and whether applied to goods with a pattern upon them or to goods that show no pattern at all, the result in either case will be immeasurably beyond anything that we at present possess. The woollen manufacture is not played out; there is a long and brilliant future for it yet in store, which will be realised when the nature and capabilities of its thread structure are better understood, and their higher excellences appreciated and applied.

PART XI

THE SELF-ACTOR MULE

TWELVE years ago I published a small work on the *Woollen Thread*, wherein it was stated that "the spinning mule was ultimately made self-acting by Richard Roberts of Manchester, whose 'quadrant winding motion' is one of the most ingenious mechanical movements that the world has ever seen." This has been fully borne out by what others have since written respecting this marvellous machine, the self-acting mule, both sectionally and in respect to the machine as a whole.

Mr. Richard Marsden, in his recent work on *Cotton Spinning*, says (pp. 229, 230): "The first self-acting mule was one of the greatest triumphs of mechanical genius that has ever been achieved, and as a display of the power of the inventive faculty in man's nature, surpassed anything accomplished up to that time. This statement hardly requires even that limited qualification, as though great advances have since been made in many branches of mechanical industry, nothing yet surpasses the spinning mule in the number and variability of its actions, the admirable concert of its parts, or the excellent results achieved by it." "The amount of inventiveness embodied in the new mule considerably exceeded that shown in the jenny or Crompton's machine, great as this was in each instance. Arkwright, as we have seen, was only to a limited extent a true inventor, being more distinguished for his ability to utilise the inventive faculty and labours of others, so that he can hardly be brought into the comparison. It must, however, be said in favour of Hargreaves and Crompton that Roberts had facilities placed within his reach by the progress of mechanical science and the improvement of machine construction—many advantages from which they were debarred. His genius was therefore comparatively unfettered, and hence achieved superior results. The value of his invention, however, can hardly be regarded as so great to the world as that of his predecessors, though the inventive skill manifested in it was superior. Hargreaves showed mankind how to do something essentially new and unthought of before; Crompton wonderfully improved upon Hargreaves's plan, both in the manner of its action and the extent of its results; Roberts made a great step forward on the labours of both."

In the above extracts Mr. Marsden presents with singular

clearness and justice the relative claims of each of the great contributors to machine spinning. Great credit is un- doubtedly due to Hargreaves for taking up spinning on the lines of the old one-thread wheel, and teaching mankind the alphabet of machine spinning on the jenny by showing them how to spin forty or fifty threads at once in the place of one. Crompton followed with the action of the jenny and the water frame combined in the mule, and Roberts brought up the rear and crowned the labours of his two predecessors, laying mankind under deep obligations to himself by making the mule work like a thing instinct with life. The industrial world owes much to this trio of great untitled men.

The head-stock of the self-actor requires to be much stronger than that of the old mule, on account of the many and various kinds of movements that have to take place within it, some of them of a wrenching character, and needing to be fixed to a firm structure, or the changes could not be effected with regularity and certainty. To attain this end many of our best machine makers build their head-stocks on bed-plates, both back and front, and these are braced together by strong stays, or are joined to the slip metals on each side of the head-stock. Stability is of the highest importance to the self-actor. Many attempts were made during a series of years, both in England, Scotland, and America, to make the mule self- acting, but comparatively little headway was made until Mr. Richard Roberts of Manchester took the matter in hand and devoted a considerable amount of time and attention to it. Mr. William Strutt, F.R.S., of Derby, son of Jedediah Strutt the partner of Arkwright, in- vented the first self-actor mule about the year 1790, but it is said that "the inferior workmanship of that

Fig. 59. —Front View of Platt Brothers' Self-Actor Mule.

day prevented the success of an invention which the skill and improvement of the present day in the construction of machinery has barely accomplished. Mr. Kelly, then of Lanark Mills, also made a self-actor mule in 1792, and in a letter to Mr. Kennedy of Manchester, dated the 8th January 1829, gives many particulars of his (Mr. Kelly's) alterations and improvements in respect to the mule : "I first applied *water-power* to the common mules in the year 1790—that is, we drove the mules by water, but put them up in the common way by applying the hand to the fly-wheel, and by placing the wheels (or mules) right and left; the spinner was thereby enabled to spin two mules in place of one. The mules at that time were generally driven by ropes, made of cotton waste, from a lying shaft in the middle of the room, and over gallows pulleys above the fly-wheels on each side of the room. The mode of driving was succeeded by belts, which was in every respect much better, and better adapted to self-acting mules. From the above date I constantly had in view the self-acting mule and trying to bring it into use ; and having got it to do very well for coarse numbers, I took out the patent in the summer of 1792. The object then was to spin with young people, like the water twist. For that purpose it was necessary that the carriage should be put up without the necessity of applying the hand to the fly-wheel. At first we used them completely self-acting in all the motions, the fly continuing to revolve, and, after receiving the full quantity of twist, the spindles stood. The guide or faller was turned down on the inside of the spindles, and the points were cleared of the thread at the same instant by the rising of a guide or inside faller (if it might be so called). When the outside guide wire or faller was moved round, or turned down to a certain point on the inside of the spindles, it then disengaged, or rather allowed a pulley, driven from the back of the belt pulley, to come into gear or action, and which gave motion to the spindles, and took in the carriage at the same time (similar to the way you assist the large mules in putting up). But in the above self-acting mule, which performed every motion, after the spindles were stopped it required about three turns of the fly-wheel to move round the faller, and put in action the above-mentioned pulley, that took in the carriage, which was a great loss of time. We therefore set aside that part of the apparatus or machinery, and allowed the mule to stop in the common way on receiving the full complement of twist; and the instant it stopped, the boy or girl, without putting their hands to the wheel, just turned the guide or faller with the hand, which instantly set in motion the spindles and took in the carriage, the cop being shaped by an inclined plane or other contrivance."

"It will naturally be asked, Why were not the self-acting mules continued in use? At first, you know, the mules were about a hundred and forty-four spindles in size, and when power was applied the spinner worked two of such; but the size of the mules rapidly increased to three hundred spindles and upwards, and two such wheels being considered a sufficient task for a man to manage, the idea of saving by spinning with boys and girls was thus superseded."

Notwithstanding that the course of improvement still went on, and one improvement after another was effected, the outcome of the whole was that very little headway was made towards bringing to perfection the idea of making the mule thoroughly and practically self-acting. Many spinners and mechanics, both in England and elsewhere, invented different motions and contrivances for the purpose of making the mule self-acting—Messrs. Eaton of Wiln, in Derbyshire; Mr. Peter Ewart of Manchester; Mr. Knowles of Manchester; Mr. Buchanan of the Catrine Works, Scotland; Mr. De Jongh of Warrington; and a Dr. Brewster of America. None of these succeeded thoroughly; the keystone of the arch was still somehow or other wanting, and could nowhere be found until Mr. Richard Roberts of Manchester took the matter up, and after applying his whole energies and great mechanical talent to the subject for some considerable time, took out his first patent in 1825; he afterwards succeeded in making further considerable improvements, and took out a second patent in 1830. Roberts's conception of applying the quadrant to winding the thread on the cop was the keystone of the arch which made all secure, and for the first time made the self-acting mule really and practicably possible. With the quadrant winding motion the foundation of the success of the self-actor mule was laid; the rest was only detail that could be worked out afterwards, and the work of completing the self-acting mule could now go forward with a certainty of arriving at a satisfactory result.

As we have already pointed out, the mule had early in its history become a mechanical spinner of great proficiency, the manual assistance necessary in working it forming no part of that which is of the most importance— *the draft*. This had become strictly mechanical one hundred years ago, and did not come upon the scene for the first time with the advent of the modern self-acting mule, but had been in existence and improved to a very high degree of perfection before the self-acting mule was generally adopted and used in our woollen mills. The vital part of good spinning—*the mechanical draft*—being already in existence when the self-actor appeared, and having reached

Fig. 60.—Whiteley's Self-Actor Mule.

a very high degree of proficiency, the self-actor could only take up the remaining movements—the backing off, the working of the fallers, the winding of the cop, and the varying of the speed of the spindles to suit the form of the cop during the winding. The most difficult and perplexing of all the movements to be obtained by mechanism was that of the variableness of the speed of the spindles that was necessary in order to meet the constantly differing diameters of the cop from the beginning of its formation to its completion; and not only that, but the motion that was required was one which would meet the variable speed that was needed in winding each individual stretch or draw of yarn, while the cop was in course of being built up, commencing with a quick movement, then diminishing in speed every instant down to the greatest diameter of the cop, then increasing from the largest diameter up to the smallest diameter, all this variation taking place during the winding on of a single draw of yarn, in addition to the variation from the commencing of the cop up to the point where it attains to its greatest diameter. This variableness in the revolutions of the spindles to accommodate the cop at all the stages of its different sizes during its formation was one of the toughest tasks to accomplish out of the whole lot, and the method devised by Roberts to meet the difficulty is one of the most ingenious inventions on record. The quadrant thus adapted and applied is one of the greatest triumphs of mechanical ingenuity the world has ever seen. The movement to be obtained, to any one that has been practically engaged in spinning, seems so complicated and difficult as to baffle indefinitely all attempts to accomplish it mechanically. Roberts's quadrant winding motion, containing the differential screw, solved the difficulty. Prior to this invention

the practically useful self-acting mule was a thing in the
far-off future, but with the quadrant winder it was an
accomplished fact.

The inclined plane, or copping motion, and other points
were one by one surmounted, and the self-actor under
Roberts took practical shape, and its success was assured.
Others since Roberts's day have made improvements, but
the self-actor foundations were undoubtedly laid by Roberts.
The quadrant winding motion alone made the self-actor

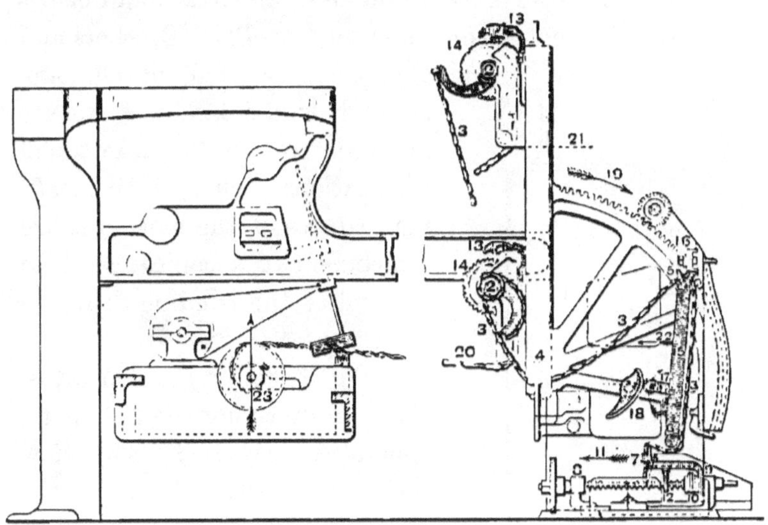

FIG. 61.—PLATT BROTHERS' PATENT NOSING MOTION.

practicable. Without this or its equivalent, making all the
other parts and motions mechanical would have been in
vain. The difference in the number of the revolutions of
the spindles between the commencement of the cop on the
bare spindles and the number required when the cop has
attained its full diameter on a woollen mule of $2\frac{1}{2}$ inch
pitch of spindle is very great; yet the quadrant provides
for this difference. A chain is attached to a nut on the

quadrant screw, and then fastened to the winding drum, round which it is coiled, and the number of revolutions of the spindles in winding on the stretch of spun yarn is regulated by the length of chain that is unwound from the winding-on drum. This winding-on drum is in the middle of the carriage, immediately under the head-stock, and is connected with the tin roller by means of click gearing. Many improvements have from time to time been introduced in working the quadrant chain, some in the shape of nosing pegs for depressing the chain, and others for deflecting it in various ways; but the most important and effective is the " patent nosing motion " of Platt Brothers and Company, Limited, of Oldham, through whose kindness we are allowed to produce sectional illustrations. One of the leading features of the motion is the attachment of a conical scroll to the end of the winding drum, as shown at 23 in Fig. 61.

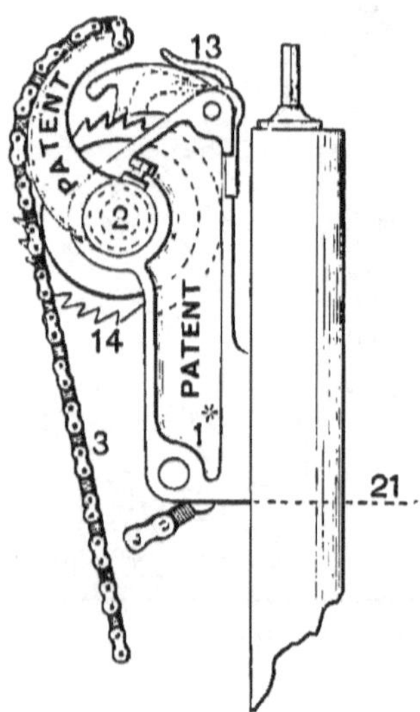

Fig. 62.—The Quadrant Nut.

" Fig. 61 shows the apparatus. The parts forming the automatic apparatus are shaded. This view shows the position in which the parts should be placed. The quadrant is shown in the position it stands in when the bowl is on the ridge of copping rail. The carriage is shown close up to the back stops—that is to say, at the end of its inward run."

" Fig. 62 shows an enlarged view of quadrant nut with the parts on it in the same position as shown in Fig. 63."

" Fig. 63 shows the position of the parts when the cop

bottom has just been completed. The carriage is shown close up to the back stops. The quadrant nut (1) is at its final position for making the rest of the cop. The arrow on the winding drum is in the position shown in Fig. 60, viz.

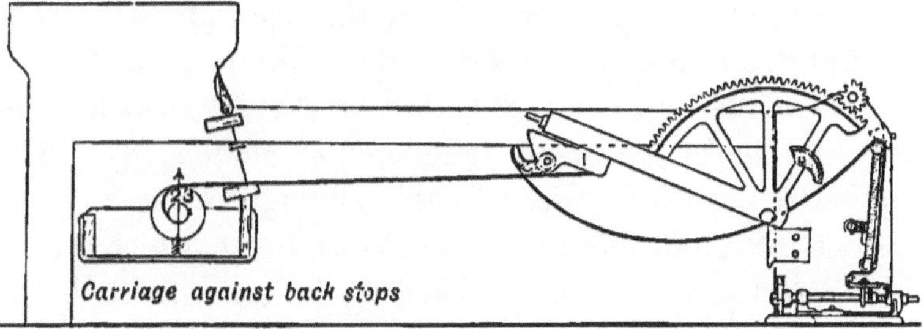

Fig. 63.—Section showing Completed Cop Bottom.

the chain has been uncoiled from the cylindrical part of the winding drum until it is on the point of uncoiling from the scroll part. The arrow crosses the drum where the circular part ends and the scroll part begins."

" Fig. 64 shows the position of the parts at the completion

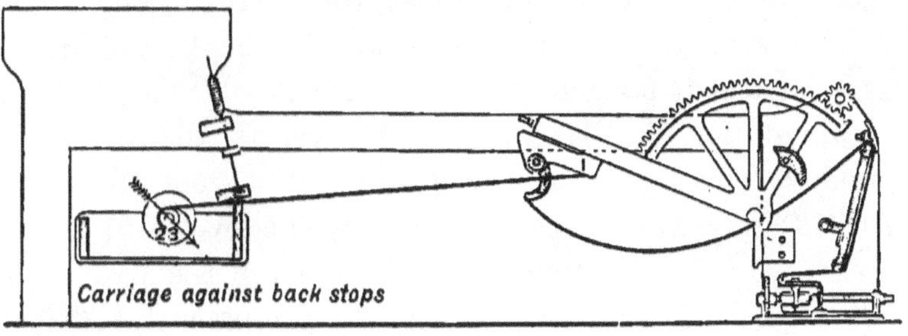

Fig. 64.—Section showing Position at the Completion of the Set of Cops.

of the set of cops. It will be observed that the winding chain is shown uncoiling from the small end of scroll."

" Fig. 61. At the bottom of the quadrant at (1) is the quadrant nut used in the nosing motion in its proper position for winding on the bare spindle. At (1*) is shown the same quadrant nut in the proper position for winding the full diameter of cop, and with the parts on it in the position they are in at the completion of the cops.   At (2) is the scroll, to which is fastened the chain (3); the other end of this chain, after passing round the pulleys (4), (5), and (6), is fastened to the sliding bracket (7), which slides on the shaper screw of the copping motion (8).   The action of the chain (3) causes this sliding bracket to press against the front nug (9) of the shaper frame, which carries the shaper screw.   This bracket (7) is so constructed that the nut (10) on the shaper screw can come into contact with it, and cause it to slide with the forward movement of the nut (10), and so gradually pull the chain (3) in the direction of the arrow (11).   The adjustable piece (12), fastened to the sliding piece (7), is for the purpose of regulating the time when the nut (10) of the shaper plate comes into contact with it.   At (13) are two detent catches, which enter into the teeth of the ratchet wheel (14).   At (15) is a lever swinging on its centre stud in the bracket (16). This bracket is fastened by a special bolt, which also serves as a bolt for holding the guard covering the quadrant pulley.   This bolt goes through the slot in front stretcher. At (17) is a swivel lever carried by the lever (15).   The lever (17) is free to turn in the direction of the arrow, but not in the contrary direction.   At (18) is a bracket, fastened to the arm of the quadrant.   This bracket is capable of pushing back the lever (15), by means of the lever (17), when the quadrant is turning in the direction of the arrow (19).   The steel end of the winding chain which goes into the quadrant nut is lengthened with another piece of steel chain about 2 feet long, shown at (20).   The curb part of the chain will therefore be shorter than usual, and no part of it passes under the anti-friction bowl in the quadrant nut ; by this means the wear of the chain in passing round the small pulley is entirely prevented."

"The copping having been properly adjusted to build a good parallel cop, it will be necessary, before the parts can be set, to mark on the quadrant the position of the quadrant nut when winding the full diameter of the cop. This may be done as shown by the line (21), which forms a continuation of the line formed by the underside of the quadrant nut. The set of cops having been finished and removed from the spindles, the quadrant nut wound to the bottom, as shown at (1), ready for a fresh start, uncoil all the winding chain from the ratchet in quadrant nut, then set the scroll (2) so that its end will point straight down, as shown in Fig. 61, when the winding chain is pulled tightly. The carriage must be run in and held against the back stops, and whilst in this position the winding chain must be cut to the proper length so that when tight it will bring the winding drum into the position in Fig. 61. It is very important that this instruction be properly carried out. Wind the quadrant nut up to the highest, as already determined and marked, the carriage remaining in the same position. Wind up the surplus chain on the ratchet by hand until the winding chain is again tight, the drum remaining in the position shown in Fig. 61. When this is done the scroll (2) ought to stand with its end upwards, as shown by (2) and (3). The copping plates should now be wound back ready for beginning a fresh set of cops. The adjustable piece (12) should be set about the middle of the slot in piece (7). The bracket (18) should be set so that when the most prominent part of the curved face is in contact with the point of lever (17) there will be a little space between the lever and its chain and the inside bead of the front stretcher. To set these, draw the carriage out slowly until the above-named bracket (18) and lever (17) are in contact, when the bracket can be set to the lever (17).

"The carriage should remain in this position whilst the following instructions are carried out: Wind the copping plates forward until they are in the position they occupy when just completing the cop bottom, viz. when the stud supporting the front end of the copping rail is just on the point of moving down the straight part of the front copping plate. The chain (3) must now be set by passing it round the little pulleys (4), (5), and (6), and down through the hole in the sliding bracket (7), and secured by a set screw. The chain must be tight in the above position, and will cause the bracket (18) and lever (17) to be in contact."

"The copping plates should be wound back against the stop, ready for beginning a fresh set of cops, and the quadrant wound to its lowest; this will cause the chain (3) to be slackened very much. Lift the detent catches (13) from the ratchet wheel and uncoil the winding chain; this will coil the chain (3) on the scroll (2), the same number of laps going on the one as comes off the other. The chain (3) will in this position be slack, and not tight as first set. The automatic apparatus being set according to the foregoing instructions, the mule may now be worked, the *modus operandi* being as follows: After the first stretch the governor motion from time to time moves the quadrant nut farther from the centre of the quadrant to adjust the winding to the increasing size of cop. By this movement of the quadrant nut the slack in the chain (3) is soon taken up. After the slack is taken up the further movement of the quadrant nut causes the chain (3) to pull the lever (15) in the direction of the arrow (22). On the return of the carriage the bracket (18), acting on the lever (17), forces it back again, and in so doing pulls a length of chain from the scroll (2), which causes it to turn on its axis and wind a length of winding chain on the ratchet (14), which would otherwise remain coiled on the winding-on drum. By this means the length of chain left on the winding-on drum at the end of each stretch will remain nearly the same until the cop has attained its full diameter, when the parts should be in the position shown in Fig. 62 (3). Up to this stage the bracket (7) will still rest against the nug (9) of shaper frame. About this time the nut (10) of shaper will come into contact with the sliding bracket (7), and gradually draw the chain (3) and lever (15) along with it, and, as before explained, draw a further length of chain from the scroll (2), wind up a further length of winding chain on the ratchet (14), and cause it to unwind from a portion of the scroll part of the winding-on drum. The repetition of this action gradually brings the scroll part of the winding drum (23) into use, until at the completion of the set of cops as much of the scroll has been brought into use as the circumstances of the case require. By causing the nut of shaper screw to act on the bracket (7), sooner or later a greater or less amount of the scroll of the winding drum may be brought into use. At the completion of the set of cops the scroll (2) should be in the position shown in Fig. 61 at (1*). Before beginning a fresh set of cops the winding chain must be uncoiled from the ratchet (14)."

Messrs. Platt Brothers and Co. have also added a "patent backing-off chain tightening motion" to their number of improvements to the self-acting mule, and describe the apparatus and its use as follows :—

"The object of the above apparatus is the automatic regulation of the backing-off chain. Fig. 64 shows the details, etc. Before explaining how the apparatus must be set, we will explain why the apparatus is required.

"When the carriage is coming out, the front or winding faller wire is generally about $1\frac{1}{4}$ inches above the spindle points. This is also the position of the parts in ordinary mules immediately preceding the operation of backing off. As is well understood, the reversion of the tin roller causes both the tin roller to uncoil the yarn from the spindles, and also brings into action the parts which pull the faller wire down. In all cases the spindles begin to uncoil before the faller wire begins to move, because the tin roller must make some little movement before the backing-off click can take hold. In addition to this, the spindles continue to uncoil the yarn during the time the faller wire is moving from its position above the spindle points until it touches the yarn. From this it will be understood that a considerable length of yarn will be uncoiled from the spindles before the faller wire can overtake the yarn. The spindles get the start very considerably. At the completion of a set of cops this difficulty or loss of motion is greatest. In the case of a cop with its nose only $\frac{3}{4}$ inch from the spindle the loss is nearly one-half, the faller wire moving as much before it touches the yarn as it does after. To overcome this difficulty it is necessary to have the backing-off chain tight, so that it may act on the faller as early as possible, and the backing-off snail is made as large as possible, so that it may get on the yarn at the earliest moment before the faller locks. The backing-off snail is specially made to suit these requirements. At the commencement of a set of cops the conditions are very much more favourable, the space passed through by the faller wire before it touches the yarn being very much less in proportion to the entire distance passed through by the wire before the faller locks than it is at the completion of the set of cops. Consequently, the backing-off chain must be slack at the beginning of the set of cops, otherwise the speed of the wire would force the yarn down the spindle faster than it would uncoil. These conditions are a compromise, but in practice it is found to work very well. The backing-off chain having been adjusted to the proper length to back off nicely at the commencement of the set of cops, it is found very desirable to gradually tighten the chain (or, as it is generally termed, 'shorten' it, by taking it up) as the cop increases in length, until at the completion of the cop the chain is almost tight. By this the backing-off all through the set can be adjusted so that it corresponds at every stage to the exact requirements of the case, and the nose of the cop is preserved in a proper condition ; neither too much yarn is uncoiled, nor too little. Next to winding the yarn properly on the cop this is the most essential condition for a good cop. Where this apparatus (which effects this automatically) is at work it is found that very much fewer noses are 'hatched,' and are therefore to a great extent free of a great drawback."

"Fig. 65. At (24) is the winding faller shaft, and at (25) is the backing-off finger; at (26) is the backing-off chain, fastened at one end of the finger (25) and at the other end to the backing-off snail (27), mounted on the tin-roller shaft (28). At (29) is the backing-off tightening chain ; one end of it is fastened to the boss of the snail (27) ; the other end is fastened to the lever (30), mounted on the centre stud (31). The other end of this lever (30*) is shown resting on an incline (32). This incline slides and rests upon a plate (33) fastened to the floor."

"Fig. 66 shows this incline and its connections *in the plan*. At (34) is the copping-plate connecting rod. An arm (32*) of the incline (32) grips the rod (34), and by means of two hoops (35) fastened to the rod (34) by set screws, the incline bracket (32) is caused to move with the motion of the rod during the formation of the cop."

"Fig. 65 shows the positions of the carriage and the

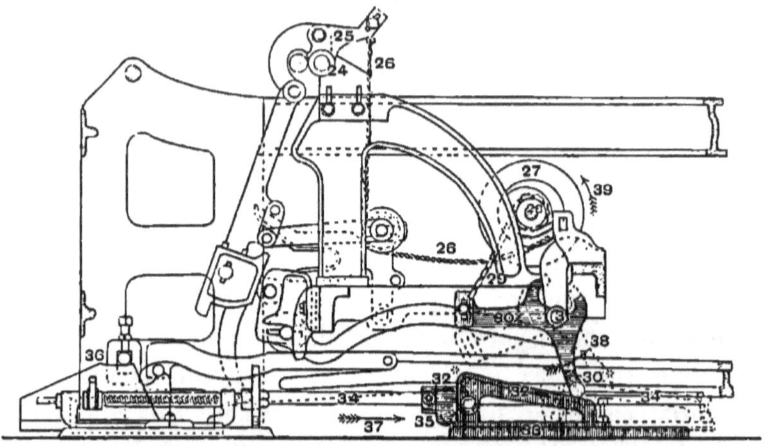

FIG. 65.—SECTION OF A SELF-ACTOR.

various parts just previous to the backing-off taking place, and at the commencement of a set of cops. The backing-off chain having been adjusted to the proper length for backing-off on the bare spindle, the copping plates (36) and the rod (34) will gradually move as the cop progresses in

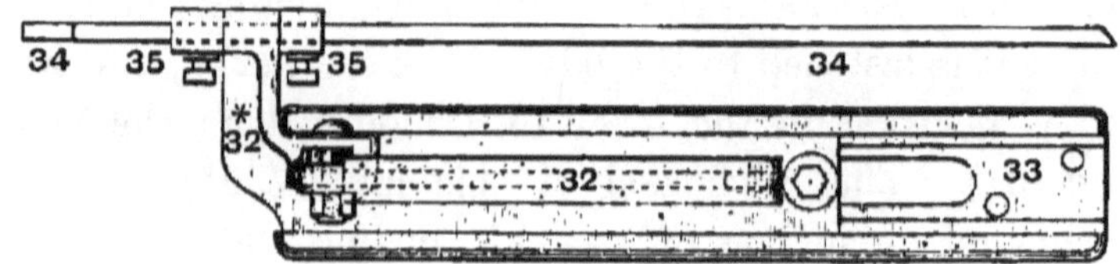

FIG. 66.—SELF-ACTOR INCLINE.

the direction of the arrow (37), and carry the incline (32) with them in the same direction. This movement gradually brings the higher parts of the incline (32) under the tail of the lever (30*), causes it to turn in the direction of the arrow (38), and pulls down the chain (29), which, acting on the snail (27), turns it in the direction of the arrow (39). This movement of the snail (27) takes up the slack of the backing-off chain (26). The incline (32) is constructed so that the inclinations can be varied to suit the particular requirements of various kinds of mules. The amount of tightening depends upon the setting of the incline; the ratio depends upon the form of the incline. By varying the form of the incline, the action on the chain can be varied to suit any circumstances. When once set the apparatus needs no further attention."

" When the copping plates are wound back for a new set of cops the incline (32) goes with them, and the backing-off chain is restored to the proper position for commencing a new set of cops."

Fig. 59 is a front view of the complete head-stock of Platt Brothers and Co.'s self-acting mule, some of the motions of which we have just noticed in detail. It will be observed that, contrary to the general usage of machine makers, the quadrant is inside the head-stock, instead of on the right hand outside.

The mule as now made includes several recent improvements. It is generally arranged to spin from condenser bobbins only, but can be made with a creel for second spinning, or for doubling or twisting, when required for that purpose. It is supplied with double-speed motion, twisting-in or drawback motion, twist motion, motion for stopping the mule, spindle-stopping motion, governing and quadrant-regulating motions, positive driving for cam shaft, patent rope-band tightening for rope taking-in motion, with two scrolls and check scroll, long copping rail, and double copping or shaper plates; automatic nosing motion, backing-off-chain tightening motion (both of these we have noticed in detail), and motion for putting the double speed on at any part of the draw or stretch. Either the single or double speed may be increased or decreased independently at pleasure. The twisting-in or drawback motion can be worked quicker or slower to suit the thickness of the yarn that is being spun, fine yarns requiring a slow speed and coarse yarns a quick speed; it can be regulated to push the carriage to a greater or less distance without affecting the speed at which it is moved inwards; it also acts as a front stop to prevent the carriage from returning whilst backing-off is taking place. All these changes are effected without the use of change wheels or pulleys.

The slubbing wheel indicates the length in inches of sliver delivered from the rollers, which are regulated without change wheels. The twist motion can be regulated to give more or less twist to the yarn, also without the use of change wheels; and the amount put in is the same at every part of the cop. The stop motion for stopping the mule for piecing-up may be engaged or disengaged from any part of the mule. The governing and quadrant regulating is effected by a variable screw, which operates during the formation of the cop bottom. The patent automatic nosing adjusts and modifies the action of the winding apparatus to suit the tapering of the spindle so as to wind a tight cop nose at every stage of its progress. By the spindle-stopping motion, the spindles, at the first portion of the carriage's outward movement, are only allowed to turn very slowly while the rollers are delivering the sliver, so as to prevent too much twist being put in before the rollers stop. The quantity of twist put in during this stage may be regulated by adjusting the friction on the roller shaft, and the time it is in action may be made longer or shorter to suit the spinning; this motion is recommended for fine spinning direct from the condenser bobbins. The roller reversing motion is applied when spinning from very short wool. The motion for putting double speed on at any part of the draw or stretch can be engaged sooner or later without affecting the amount of twist put into the yarn or the relative velocities of any portion of the mechanism. The coupling shaft connecting and driving the rollers on either side of the head-stock can be removed, if required, without disturbing the rollers, and also the shaft driving the tin rollers on each side of the carriage square. This plan offers great facility in changing the pulleys, wheels, etc.

The mule is adapted to spin warp or weft, either upon the bare spindles or upon bobbins, and it can be easily changed from one to the other, having special arrangement of copping plates to suit both bare spindle and bobbins.

The change wheels are of large size, and give minute changes. The square or iron portion of the carriage under the head-stock carries all the mechanism which connects the carriage on both sides of the head-stock, and is supported by two long slips on the foundation plates, which support the head-stock.

The whole of the self-acting parts are self-contained, and do not depend on the woodwork in any way, but, instead, support it.

The rim shaft is case-hardened, and the general construction secures an easy and accurate adjustment of the various parts of the machine combined with great steadiness and durability.

Fig. 60 is a representation of the head-stock of a self-acting mule by Wm. Whiteley and Son, Huddersfield, for which the makers claim the following advantages—

This mule is of an entirely new construction, and has been carefully designed so as to allow every motion to be worked in the simplest manner possible, at the same time with accuracy and certainty, and easy of access for changes. The change wheels for both draw and twist are so conveniently arranged that the spinner can change any wheel without having to disturb any other casting, as is the case in some mules.

The following are some of the principal features in this mule :—

1. The head-stock is built on two foundations or bed-plates, and secured back and front with the middle carriage slips, making the head-stock as firm as if it were one solid piece, at the same time making the carriage middle perfectly accurate and true with the head-stock.

2. In the driving it is direct or parallel with the main shaft, which usually runs lengthwise with the mule, thereby saving the expense of cross shafts and the noise of the bevel gear.

3. The fast and slow speed of the spindles is got by two diameters of rims fixed on the rim shaft; each rim is entirely independent of the other, and can be changed to get any speed required, which is a great advantage over the one-rim principle, which takes down the fast speed when making the slow speed slower to suit the drawing of the yarn.

4. The drawing-out, taking-in, and backing-off, and strap-guide are changed by a cam, and every change is timed, so that no two opposite motions can be in at the same time, because when it puts one in it takes another out, and thereby prevents breakages.

5. The jacking-up or drawback motion can be made to jack-up from nothing to 8 inches, and at any speed

required. The carriage can also stand any length of time before the jacking-up motion comes into operation, or after, to allow the yarn to set.

6. The cop shaper being one of the most important points (after the draft) in a self-acting mule, we wish to call special attention to the improved shaper, which will make any shape required, and is not liable, as the old kind, to get out of shape; the same shaper will also do for bobbins in addition to cops, without any change of rail, by simply adjusting the gauges or copping rails.

7. The relieving motion for putting on the friction backing-off is a simple connection to the intermediate or back shaft, which the minder can adjust any time without stopping the mule, thus allowing the rim band to slacken speed before the reverse motion (that is, the backing-off) can get into gear, thereby preventing those jerks on the rim band, and causing them to wear much longer.

8. The mule is very firmly put together, and is constructed on the interchangeable principle. All quick-running shafts are case-hardened, all the bearings are bushed, and all change-wheels are cut out of the solid by special machinery. This mule is the product of a woollen district, and in its present form is a thoroughly good working, useful, and reliable machine.

The self-acting mule, in receiving the two motions of the Messrs. Platt Brothers and Co., previously detailed, received an addition to its efficiency that enabled it to make a great forward movement, and, as an automatic machine, enabled it to attain to comparative perfection. Roberts's quadrant motion laid the foundation for the success of the self-actor mule. Details of minor importance have been added from time to time since his day, in the way of the nose-peg and the swing lever for acting on and deflecting chain ; but the two motions above referred to and detailed have brought the self-actor fairly within sight of practical working perfection, and made the machine available for the spinning of a great variety of work, both in wool and cotton. On Roberts's foundation other men could see their way to build, but until this foundation was laid little advance was made towards making the mule self-acting or automatic. As Stephenson gave working form to the locomotive, so in like manner Roberts gave working form to the self-acting mule. Providing the spindle had been of uniform diameter, the Roberts's quadrant would at once have covered the whole ground of the winding of the cop, but the varying diameter of the spindle required an acceleration of the speed of the spindle during the last few inches of the running in of the carriage, and the required acceleration has now been obtained by means of the patent nosing motion of Messrs. Platt Brothers and Co. This firm's patent nosing motion supplements the action of the quadrant at the precise point where it was needed, especially in the cotton trade, and keeps the nose of the cop firm and neat. Just as much, or as little, of this accelerated speed of the spindle can be made use of as is found necessary to keep the nose of the cop right in the winding. The other motion, as we have seen, has reference to the backing-off and the manipulation of the fallers, and the assuring of the more certain and uniform action of both these sections of the machine, so as to prevent snarls by undue slackness of thread during backing-off, and so forth.

For most kinds of woollen spinning the quadrant, as Roberts left it, was quite sufficient for all winding purposes, as the taper of the spindle is not as much in the woollen mule as in most cotton mules, and the

thread is elastic and will "give and take" in a manner
that the cotton will not; but with the two motions we
have described to supplement the action of the quadrant
the winding can be adjusted with the greatest nicety,
and make the working of the self-actor practically
complete. The Figs. 59 and 60 may be taken as a
fair general representation of the modern woollen
spinning mule with the most recent important improvements, by which, with the suggested alterations and
additions in respect to the draft-scroll previously
pointed out, spinning may be carried to a high degree
of perfection, as the two motions noticed of Messrs. Platt
Brothers and Co., in addition to those previously in use,
make the working of the self-actor mule practically complete; and in closing our remarks upon the machine we can
only repeat our warning that we must not expect that
it will improve the quality of our spinning, and that
we shall immediately make a great bound forwards,
because the quality of the spinning depends solely upon
the draft, and that has been purely self-acting and
strictly mechanical almost from the first existence of the
mule, or at least ever since the invention of the
draft-scroll, over one hundred years ago. The self-acting
is only the making of the other portions of the machine
—the secondary portions—self-acting that were not so
before in the old mule. Because we succeed in making
the secondary motions of a machine automatic or self-
acting, it does not necessarily follow that we shall greatly
improve the quality of its production; we shall therefore
do well to moderate our expectations as to the quality
of our spinning being much improved by the self-actor
mule; nevertheless, this much may now be said, that
our spinning machinery is at this day much more com-

plcte and perfect than our knowledge of spinning, and that we must look for the advance in other directions than in our spinning machinery, which is much ahead of the knowledge of using it properly, and up to its present capabilities, judging by some recent uses to which the woollen mule has been applied, as we in the woollen manufacture seem to have completely lost sight of the intent and object for which spinning was instituted as one of the processes of manufacturing woollen fabrics, or we should not have displaced 95 per cent of our spinning in order to make way for such a machine as the condenser.   What we want is an " EXPANDER."

PART XII

MISCELLANEOUS ITEMS

Skeins or Counts in Woollen.—Another matter in connection with woollen spinning is the " skeins " or counts by which the thickness of the yarn is reckoned.   A woollen skein is a hank containing 1520 yards.   The very name implies that a " skein " is a hank of some kind or other, and the number of skeins by which any given woollen yarn is called is the number of skein hanks that it takes to weigh a " wartern," consisting of 6 lbs. avoirdupois.   If a certain yarn is called a 20-skeins yarn, then it means that the yarn has been spun to such a degree of fineness that it takes twenty hanks or skeins, each containing 1520 yards, to weigh 6 lbs., or a " wartern."   The " skein " and the " wartern " are both very old woollen standards by which the thickness or degree of fineness to which woollen yarns are spun are tested, and are the basis on which woollen yarns are numbered.

Just as in worsted the hank contains 560 yards, and the counts are the number of hanks, each containing 560 yards, that it takes to weigh one pound avoirdupois, so, in its more ancient but kindred industry, the woollen, it is the number of skeins or hanks in 6 lbs. that is the basis on which the numbering is founded. This old plan of numbering woollen yarns dates so far back that comparatively few people know the basis on which it rests, but it should be the habit of the student of woollen spinning never to content himself with just scratching the surface of things as he goes along, if it be only a secondary matter like the counts in spinning, but to dig down to the basis on which the upper structure rests. Things are more easily mastered when studied in their simple original elementary form than later on, as we instanced when referring to carding and other matters.

There is an erroneous notion afloat that the number of skeins to which a yarn is spun means the number of yards that it takes to weigh a dram ; but a yard is not a skein ; the very name (skein) implies that the thing in question is a hank of some kind or other that has been wound on a reel. This is one of those " short cuts " that the student should never rest satisfied with, but should strive to lay hold of the foundation principle with which it is connected. The woollen skein contains 1520 yards, and the " wartern " of 6 lbs. contains 1536 drams, so that there are not as many yards on the one hand as there are drams on the other, but the numbers come sufficiently near to each other as to constitute a " short cut " for testing the thickness, or degree of fineness, to which yarns are spun, as it comes within a small fraction of the truth, the error being only one ninety-fifth of one, and a yarn tested on this basis does not get so much as one number astray until it reaches 95. A yarn said to be 95 skeins per wartern on the "short cut" system of measurement is really 96 skeins, as there is one ninety-fifth fewer yards in a woollen skein than there are drams in a wartern of 6 lbs. The use of this short method of testing the counts of woollen yarn has now become so common that many people do not know anything to the contrary of a yard to a dram and a skein to the wartern being any other than expressions that mean one and the same thing. Such, however, is not the fact.

While dealing with skeins a further matter may be mentioned, and that is an easy and ready method of ascertaining what any worsted number represents in skeins. Take the worsted number and multiply by $2\frac{1}{5}$, and the product will tell us what the worsted number represents in woollen skeins. A 30s. worsted is equal in thickness to 66 skeins woollen, and a 48s. worsted is equal in thickness to $105\frac{1}{2}$ skeins woollen. This, like other "short cuts," is not absolutely correct, but is deemed near enough to give a general idea, as it comes within a fraction of the truth, and is sufficiently near for many purposes.

Twine in Woollen.—In connection with spinning there are two twines in woollen. One is called "cross-band," and the other is called "open-band." If a sliver be taken, and one end be held fast between the finger and thumb of the left hand while the finger and thumb of the right hand twine the other end, if, on examining the twined sliver, the folds of the twine incline from right to left, the yarn is called "cross-band," and if the folds of the twine incline from left to right it is called "open-band."

The origin of these two names or terms, "cross-band" and "open-band," dates back to the days of the one-thread spinning wheel. When the string or band that went round the *rim* or circumference of the wheel went round the whorl or wharl in a direct line, the twine was then called " open-band "; but if the string or band from the top side of the *rim* went down to the under side of the spindle wharl, and the string from the top side of the wharl went down to the under side of the *rim*, then the twine of the yarn was called " cross-band," because the band or string was crossed in the intervening space between the rim and the spindle, —hence the term " cross-band " twine. When the band ran open between the wheel and the spindle the twine was called " open-band." The reason why the spinner has sometimes to spin the yarn " open-band " twine and at other times " cross-band " is in order to meet the requirements of the cloth for which the yarn for the time being is intended to be applied. In January 1877 I contributed a paragraph to the *Textile Manufacturer* on the subject of the two twines, from which the following extract is taken : " In the production of all fancy cloths, whether in self-colours or in various colours, design or pattern has to be aimed at, and this involves a certain amount of sharply-defined outline ; and in order to obtain and retain this sharpness of outline, not only must the warp and weft threads cross each other at right angles, but the folds of the twine of the warp and the weft threads must cross each other at right angles also, to enable the threads to retain their individuality in the after operations of ' fulling,' ' finishing,' etc. In using warp and weft spun the same way of twine, the folds of twine in the warp and weft cross each other in the cloth at right angles, thereby removing, as far as it is possible to remove, the liability of the fibres of the threads, and the threads themselves, to mingle bodily in the fulling, etc. Then

turn to the opposite of this in the manufacture of plain
cloths, where it is requisite to hide as much as possible all
outline of the 'make' of the cloth.   To obtain this result
the weft requires to be twined the opposite way of the
warp, in order to afford the greatest facility for the fibres
mingling and felting and forming one homogeneous mass,
hiding every vestige of outline of the 'make' or framework of the fabric, and presenting a close, even, compact
surface for the 'finisher' to operate upon.   In the fancy
cloth it is requisite to preserve as much as possible the
individuality of the threads for the sake of the pattern,
and in the plain cloth it is requisite to lose it as quickly as
possible for the sake of obtaining closeness of face and cover ;
and such being the case, we must have the folds of the
twine of the weft to meet, or coincide with, those of the
warp, so as to mingle quickly in the fulling, and not to
cross at right angles, as in the fancy cloth.   The one rule
gives distinctness of pattern, and the other closeness of face
by sacrificing the pattern.   By using opposite twines in a
fancy cloth, closeness and evenness of face is obtained, as
in a plain cloth, but pattern is sacrificed in doing it.   The
one rule is applicable to the fancy manufacture, the other
to the plain manufacture."

Fancy Woollen Twists. — There is a variety of twists
producible in connection with the spinning for use in the
different branches of the woollen manufacture.   There are
the knickerbocker, the clouded, and the diamond (see
Fig. 67).

I took out a patent (No. 3069) for the production
of the knickerbocker twist in a much less tedious and
expensive manner than had previously been accomplished.   This fancy twist had been produced in France
by a very slow and tedious process at about 5s. per lb.,

but by the above patent process the same counts of yarns
could be converted into knickerbocker twist at 3d. per lb.
Very fine counts, where very fine silks were employed in
forming the spot, would frequently cost 6d. per lb. to twist,

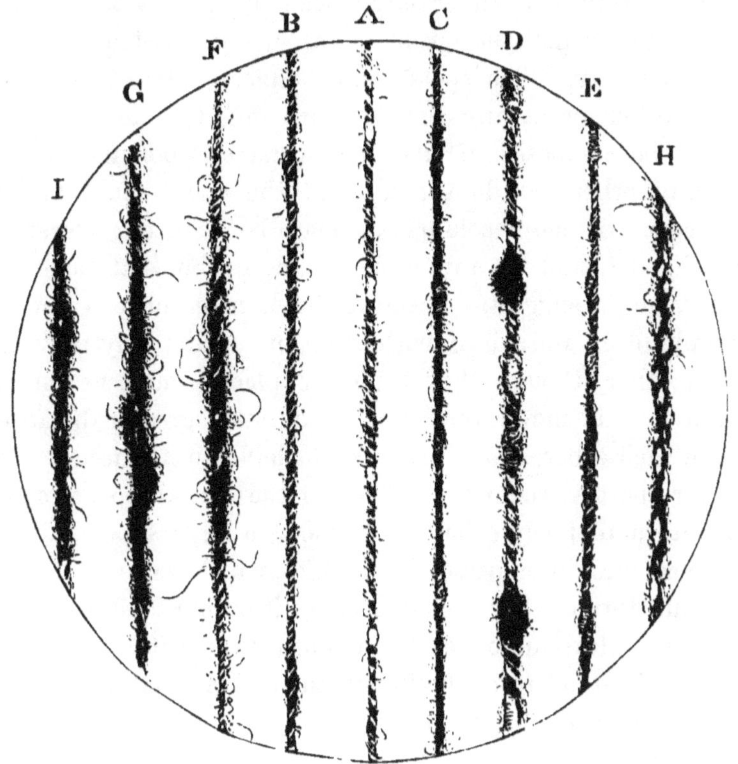

Fig. 67.—Varieties of Twists.

that is to say, for the cost of twisting the yarns alone,
in addition to the cost of the single yarns. The patent
was taken out in November 1866, and was available for
application either to the woollen spinning mule, or to the
twisting frame, but it worked for some time on the mule

first, though the specification included the twisting frame as well as the mule.

The knickerbocker twist was produced by taking two single threads, either of the same colour or of contrasting colours, to form the ground or foundation thread, and then super-adding a small or large bead or spot by the patent apparatus while the ground was being twisted in the ordinary way. The spots could be placed almost close to each other or at any distance that might be desired, and could be composed of silk, fine worsted, woollen, cotton, or any other textile thread, and the spots could be of alternate or more colours, as may be seen on reference to Fig. 67, and in any order of succession that taste or fancy might suggest. A shows the knickerbocker twist in simple black and white, with the spots moderately near to each other, though they could be placed much nearer if desired. B shows the spots at a much greater distance from each other, though they could have been placed twice as far apart if required. C shows the knickerbocker spot in alternate fancy colours. D shows a large spot in alternate colours, interspersed with white, which was used in the manufacture of shawls, and E shows it in four fancy colours, placed in close succession to each other, so that the student will understand from the illustrations given that the spots could be varied in size and distance to suit any kind of manufacture. Besides the above illustrations showing the spots in round beads, they were produced in a variety of long trailing forms, say of a quarter of an inch length and onwards towards 2 inches in length for ladies' dress goods and other purposes.

When using a good quality of silk some very pretty effects were got, both in the round bead spot and in the long spot, and both were extensively used for the checking and striping of ladies' dress and other goods, and many of the effects are to be seen season after season either in one form or another down to the present time. The present spring season shows many of these effects quite as fresh as ever in many dress styles.

Though every season for nearly thirty years this spotted effect has been in use in a variety of forms, yet, in consequence of "Our Cruel Patent Laws," it brought no advantage to myself beyond landing me in a lawsuit, with all its expenses and anxieties.

Clouded Fancy Twist.—The other kind of fancy twist that has been referred to is the clouded one. It is produced on the mule, and was suggested to the writer by observing the action of the twine in rushing in and taking possession of the thin place or "twit" which occurs when there is a slip in the thread during the drawing in the spinning. We stated when describing the operation of the woollen draft that sometimes when the twine overpowers the draft during drawing, the fibres, instead of gradually reducing in bulk, suddenly slip in a thread here and there, and make a very thin place, or what is technically called a "twit," into which the twine instantly rushes, and stops all further drawing for the time being until the remaining portions of the threads get reduced in substance down to an equality with the thin place or "twit." The "twit," then, is the basis of this kind of clouded yarn, and the "twit" is artificially produced, so as to allow the twine to have the opportunity of rushing in. In addition to the ordinary set of winding-in rollers of the woollen mule, another set of rollers have to be placed immediately behind the ordinary set common to the woollen mules. When the mule is at work upon this kind of twist the ordinary set of rollers run the whole of the time that the carriage is receding, but the extra set of rollers that are placed behind the ordinary set work intermittently ; say, for instance, that they move at the same speed, when moving, as the ordinary rollers for an inch, and then stand still for such a space of time as it takes for the ordinary rollers to wind in an inch in length of thread, then they recommence moving again, and so on, running and stopping alternately. Bearing in mind this action of the rollers, notice in the next place how the slivers and threads are arranged. There has to be a separate set of guide wires immediately behind each set of rollers, and the two sets of rollers have to be as near each other as they can be conveniently placed after allowing room for the guide wires. Then when the rollers and the guide wires have all been arranged, the spinner takes a spun thread, twined to a full warp twist for each spindle, and threads it through the front rollers only, and this spun thread forms the ground or foundation of the clouded twist. The condenser bobbins are next taken and put into their places and the sliver drawn through both sets of rollers. It will be borne in mind that the spun thread was only threaded through the front ones, having been brought over the tops of the back set of rollers and threaded through the front or ordinary set. This arrangement being kept in mind, it will be seen that when the mule commences running, the twine of the spindles being the contrary way to that of the spun ground thread, as if for an ordinary twist, the ordinary or front set of rollers run continuously ; but the back set of rollers run only for an inch and then stop, and as the soft sliver from the condenser bobbins is threaded or drawn through both sets of rollers, it follows as a natural consequence that the front rollers running continuously gradually stretch the soft sliver to the point of breakage ; but just as this point is being approached the back rollers commence to run again just

before the soft sliver is actually broken off, and in this way
the back rollers run and stand alternately inch by inch
when the mule is at work, and the front or ordinary set of
rollers run continuously, as in common twisting work.   The
effect of the double set of rollers acting in this way is that
the soft sliver part of the clouded twist issues from the
front rollers in thick and thin portions alternately, and
produces a twist thread of thick and thin portions in
alternating succession (see F in Fig. 67).   The mule works
in every respect as it would when engaged upon any
ordinary twisting or doubling work, excepting as regards
the action of the extra set of rollers that have been placed
at the back of the ordinary set.   There is nothing whatever
between the ordinary rollers and the point of the spindles
to control in any way the twine, and yet when the twist is
examined the turns of twist in the smallest parts where
the sliver has been almost stretched asunder are as many
as twenty per inch, while the thickest parts of twist
thread will be found to be twined no finer than at the
rate of about three turns per inch.   This arises from the
fact that we have purposely produced an artificial "twit"
every alternate inch, and the twine rushed into the small or
"twitty" place, and the finished twist varies from twenty turns
per inch in the smallest places to about three turns per inch
in the thickest places.   The ground thread, which only goes
through the front or ordinary set of rollers, is of uniform
thickness throughout, but it will have been seen that the
soft sliver has been nearly broken asunder every alternate
inch by the stoppage of the back set of rollers; and if the
twist is only a simple black and white, the white shows
in three times as much bulk as the black ground thread
in some parts, while in others they appear about equal—
hence the twist is cloudy.   It is capable of an almost

indefinite variation. The soft sliver may have a soft roving of any kind of fancy colour run in along with it, or the ground thread may be of two or more colours (see I in Fig. 67, and G and H) where the cloud twist is cross-twisted with threads of fancy colours. Having been once swindled by "Our Cruel Patent Laws" no attempt was made to patent this invention.

THE END

Printed by R. & R. CLARK, *Edinburgh*